Methods of Theoretical Physics and Their Applications to Biopolymer Science

METHODS OF THEORETICAL PHYSICS AND THEIR APPLICATIONS TO BIOPOLYMER SCIENCE

L.V. YAKUSHEVICH

NOVA SCIENCE PUBLISHERS, INC.
COMMACK

Art Director: Maria Ester Hawrys
Assistant Director: Elenor Kallberg
Graphics: Denise Dieterich , Kerri Pfister
Erika Cassatti and Barbara Minerd
Manuscript Coordinator: Roseann Pena
Book Production: Tammy Sauter, Benjamin Fung
Circulation: Irene Kwartiroff and Annette Hellinger

Cover Design: Kerri Pfister

Library of Congress Cataloging–in–Publication Data available upon request

ISBN 1-56072-246-0

6080 Jericho Turnpike, Suite 207
Commack, New York 11725
Tele. 516-499-3103 Fax 516-499-3146
E Mail Novasci1@aol.com

Printed in the United States of America

The author dedicates this book to the memory of her first teacher and friend Dr. Andrew Blokhin

CONTENTS

ABSTRACT

This monograph provides an introduction to physical theoretical methods in biopolymer science. Primary attention is paid to the methods of quantum statistical physics and nonlinear wave theory, which permits to calculate both the equilibrium macroscopic properties of biomolecules and various dynamic parameters characterizing the internal mobility of biopolymers. Two classes of biopolymers are considered in detail – alpha-helical proteins and DNA.

The book contains many applications of the methods. Most of them are based on the original investigations of the author.

The book is intended for physicists, mathematicians and biologists. It can be used also as part of a course of biophysics for the last-year students and post-graduates.

PREFACE

Now, at a time when the passion for computers and computer-related methods grows general and common, I would like to call attention of my readers to another group of scientific methods usually called analytical ones. Though this group of methods is in the shadow there is no doubt about the great importance and potentialities of them. The analytical methods have, perhaps, only one serious defect: in most cases they lead only to qualitative results, and the theoreticians using analytical methods, hardly can expect a good quantitative coincidence of theoretical with experimental data.

Analytical methods are a basis of modern theoretical physics. Those of them which are presented in the book have been successfully applied in various fields: nucleic physics, solid state physics, hydrodynamics, optics, plasma physics and others. An important feature of the monograph is that the analytical methods, which have been well worked out and tested in many physical problems, are applied to a new and unusual object – biopolymers. Such a synthesis of two different fields of science, theoretical physics and molecular biology, in one book became possible due to an event which occurred several years ago, when after graduation from the Theoretical Chair of the Physical Department of the Moscow University directed by N. N. Bogolyubov, I got the opportunity to work at the Institute of Biological Physics of the Academy of Sciences of the USSR. The coincidence of these two circumstances determined the character and direction of my further scientific interests.

According to the models considered, the book divides into three parts: molecular chains which are assumed to be simplified models of biopolymers, alpha-helical proteins and DNA. According to the methods used the material of the book could be divided into two parts: the methods of quantum statistical physics and those of the nonlinear wave theory. One more division is invitingly present in the book: linear and nonlinear methods of mathematical physics.

Taking into account the interests of a wide range of readers, I begin the monograph with a brief digression into molecular biology which includes the description of the structure and dynamic properties of proteins and nucleic acids. Two classes of biopolymers, alpha-helical proteins and DNA, are considered in detail. The next part deals with the main definitions and methods of quantum statistical theory. Applications of the theory to studying the excited states of simplified models of biopolymers - molecular chains - are described in Part 3. Part 4 is concerned with the methods of nonlinear wave theory which is now undergoing fast development. Various applications considered in Part 5 are an interesting attempt to spread the ideas and methods of nonlinear wave theory to molecular biology. The conclusive Part 6 contains a short description of methods and applications which have not been included in the book. It also contains a brief discussion of the perspectives of further applications of the methods of theoretical physics to biopolymers.

The material of the book is given in a rather completed form form. However, the reader is assumed to be familiar with the main elements of physical theory including classical mechanics, quantum mechanics and statistical physics.

Only several years ago the difference between the methods and subjects of theoretical physics and those of molecular biology was so considerable that any attempts to unify them seemed to be bound to fail. But enthusiasm of the youth, remarkable examples demonstrated by Schrödinger [1] and Szent-Györgyi [2], and the support of my chief prof. Kitaigorodskii A. I. pushed and helped me to take this risky path and to present you now the results in this book, which I hope the readers will find interesting and useful.

I would like to express my gratitude to the colleagues I. B. Kuzminikh, V. I. Abakumova, S. V. Sidorova, M. S. Il'ina and G. V. Nekrasova for their constant support and help in preparing the manuscript for publication. I thank also Dr. G. Gaeta for reading the manuscript and valuable corrections.

PART 1. INTRODUCTION.

According to Pauling [3], molecular biology is a range of sciences dealing with the fine structure of living matter and the processes of life activity at the molecular level. So, the main subject of molecular biology is biomolecules, their structure, dynamics and functioning. The latter includes interaction of biomolecules with ligands, formation of complexes, the role of biomolecules in biochemical cycles, in regulation of gene expression and so on. In this part we present brief information about the structure and dynamics of biomolecules. Two classes of them are considered in detail: alpha-helical proteins and DNA. A more detailed information about these and other biomolecules and the principles of their functioning can be found in the books of Volkenstein [4], Pauling and Pauling [3], Stryer [5], Cantor and Schimmel [6], Saenger [7] and others.

§1. Proteins.

Among the great number of different molecules taking part in processes of life, proteins and nucleic acids are of particular significance. They play an important role in the structural organization of cell and tissues, in the regulation of chemical processes of vital importance, in conservation and transmission of the genetic information of heredity. In this paragraph a short information about proteins is given. The properties of alpha-helical proteins are considered in detail.

From the chemical point of view, proteins are composed of nitrogen (about 16%), carbon, hydrogen, oxygen and other elements such as sulphur, phosphorus, iron and copper. The main biological function of proteins is to build the structural elements of organisms and to provide the energy necessary for life processes [8]. According to their functional properties, proteins can be divided into several groups. Enzymes are proteins that catalyze biochemical reactions. Transport proteins facilitate the movement of molecular oxygen and other essential compounds to the sites of their utilization. Antibodies are proteins that bind to and neutralize foreign materials that may be harmful to the organism. Some proteins are responsible for maintaining the structures of cells, organs, and organisms. Other proteins play an essential role in genetic expression, nerve conduction and other biological processes.

Proteins are rather large molecules. In order to compare the size of protein molecules with that of usual molecules occurring in organic and inorganic chemistry, we present here the chemical formula of the molecule of haemoglobin, a pigment providing the red color of blood,

$$C_{3032}H_{4816}O_{872}N_{780}S_{8}Fe_{4} \qquad (1-1)$$

Every protein molecule consists of several simple components - amino acids. 20 amino acids are known in all. Their names and formulas are given in Table 1. So the proteins of all living organisms (from bacteria to the human) are constructed of the same collection of 20 amino acids. This universal protein "alphabet" has existed for about 2 milliard years.

In proteins, amino acids are linked with each other by peptide bonds, forming long chains named polypeptide chains (Fig. 1). According to the generally accepted terminology, amino acids that are part of the polypeptide chain are named amino acid residues. As can be seen from Fig. 1, a polypeptide chain has two parts: a) a regular part consisting of repeating elements and forming the so-called main chain or "skeleton", and b) a variable part consisting of amino acid residues ($R_1, R_2, R_3 \ldots$) which alternate in a special manner and form the so-called side chains. The sequence of amino acid residues in any protein is unique and completely determined by genes.

Molecular biologists relate the existence of the amino acid residues sequence with the notion of primary protein structure. The introduction of this term as well as other terms such as secondary, tertiary and quaternary protein structures, can be explained as an attempt to distinguish different levels of structural organization of the rather complex "architecture" of proteins. The first level is the primary structure, that is the sequence of amino acid residues in the polypeptide chain. The existence of the next levels arises from the fact that polypeptide chains of proteins are not planar. They are usually folded to form some configuration in three-dimensional space.

The next level (or the secondary protein structure) is due the folding of polypeptide chain in a helix or in some other regular configuration. The secondary protein structure can be defined as a structure formed by steric interactions between amino acid residues being localized not very far from each other in the sequence. Sometimes these interactions have a regular character, that gives rise to periodical structures. Alpha-helix, betta-layer, and collagen helix are examples of regular secondary structures.

The third level is called a <u>tertiary</u> structure. This structure is formed by steric interactions between the residues which are rather far from each other in polypeptide chain. At last, proteins containing several polypeptide chains have one more level of structural organization: <u>quaternary</u> structure. This level is completely determined by the way of packing polypeptide chains.

Let us consider in detail the alpha-helical protein structure described by Pauling and co-authors [9,10] in 1950.

1. Alpha-helix look like a rod (Fig.2) with the internal part consisting of the tightly twisted main chain and the side chains placed on the helix and directed outside (Fig.3).
2. Stability of the helix is provided by hydrogen bonds between NH- and CO-groups of the main chain (Fig.1).
3. Projection of any residue on the helix axis is placed at a distance of 1,5 Å from the projections of neighboring ones. The angle between them is equal to 100°, that is, we have about 3,6 residues per turn. The helix pitch is about 5,4 Å, the diameter is about 5 Å. The length of the helix is usually less than 40 Å.

In principle, the alpha-helix can be both right-handed and left-handed. However, in reality all known alpha-helical protein structures are twisted in the same direction: clockwise; that is, they are right-handed.

Two or more alpha helices can be wound around each other to produce a superhelix. Such a structure is found in many proteins: myosin (Fig.4), keratin of hair and other fibrous proteins. Spiralized "ropes" of these proteins can form bundles of fibers and play an important mechanical role.

§2. Nucleic acids.

Nucleic acids are complex chemical compounds that contain carbon, oxygen, hydrogen, nitrogen and phosphorus. The compounds derived their name from two reasons: all of them have acidic properties, and they were first identified in the cell nuclei. In a first time it was suggested that there are only two types of nucleic acids. The first which contains the sugar ribose was named ribonucleic acid (RNA). The second which contains deoxyribose, was named deoxyribonucleic acid (DNA). Later many different types of DNA and RNA molecules were found. They differ from each other by the details of the structure and by the role they play in metabolism.

The main function of nucleic acids is to carry the information necessary for protein synthesis [8]. DNA molecules transmit the genetic information of heredity from one generation to the next. RNA molecules play a vital role in the translation of the DNA sequence of each gene into amino acid sequence of the corresponding genes.

In this paragraph we shall describe in detail the structure and properties of only one class of nucleic acids: molecules of DNA.

It is known that DNA molecules occur in chromosomes of cell nucleus and, in much smaller quantity, in mitochondria and chloroplasts. The DNA molecules occurring in natural conditions are long thread-like molecules. The microphotography of a DNA molecules is shown in Fig. 6. Sizes of some different DNA molecules are given in Table 2. It is interesting to compare the sizes of DNA molecules with those of the proteins considered in the previous paragraph. The diameter of the myoglobin molecule, for example, is only 65 Å (See Fig. 5), and the length of the collagen molecule, which is one of the longest proteins, is equal to 3000 Å. Comparing these data with the data presented in Table 2 we can conclude that the length of the DNA molecule considerably exceeds the lengths of polypeptide chains of the protein molecules (some times it is longer by several orders).

Concerning the chemical composition of DNA we can say the following. The molecule is made up of many monomers called nucleotides (Fig. 7), each consisting of phosphate, 5-carbonic sugar (pentose) and nitrous base. The bases being part of nucleotides are of four different types. Two of them - adenine (A) and guanine (G) - are purines, and the other two - thymine (T) and cytosine (C) - are pyrimidines (Fig. 8). Nucleotides, linked to one another by the etheric bonds (the sugar of one nucleotide is linked to the phosphate of the next nucleotide) form a polynucleotide chain, the backbone of which consists of regularly alternating sugar and phosphate groups. The bases bound to sugar groups form a sequence along the chain. It is this sequence that determines the so-called primary structure of DNA (the reader can compare this definition with that of the primary structure of proteins).

The DNA molecule consists of two polynucleotide chains. The way by which they are held together, i.e. the principles of the formation of the secondary structure was found by Franklin, Watson, Crick, Wilkins and coauthors [11-14] in 1953. The main features of the structure can be formulated briefly in the following way.

1. Two polynucleotide chains are wound around a common axis to produce a double helix (Fig. 9).
2. The diameter of the helix is 20 Å. The adjacent bases are 3,4 Å apart along the axis and rotated 36° with respect to one another. So we have 10 nucleotides per one full turn of the helix which corresponds to a length of 34 Å.
3. Bases are located inside and phosphates and deoxyribose residues outside the double helix.
4. The two polynucleotide chains are held together by hydrogen bonds between the bases as shown schematically in Fig. 10. The bases are joined together in pairs, a single base from one chain being hydrogen-bonded to a single base from the other.
5. Only certain pairs of bases are possible. One member of the pair must be a purine and the other a pyrimidine. Possible pairs of bases are shown in Fig. 11. Due to specific pairing, one polynucleotide chain is complementary to the other.

The results of contemporary investigations of the DNA show that its structure is not rigid or fixed. On the contrary the molecule possesses a considerable internal mobility. This becomes apparent if we recall the polymorphic properties of the DNA molecule, that is the existence of many different forms of DNA. The most known of them are the A-form, B-form and Z-form. The parameters of the double helix presented above refer only to the B-form. Another manifestation of the conformational flexibility of the DNA molecule is the existence of superhelical DNA structures, which are possible due to the ability of the double helix axis to bend and to coil.

§3. Simple physical models of biomolecules.

In the previous paragraphs a brief description of the structure and properties of two classes of biomolecules – alpha-helical proteins and DNA – has been given. As it appears from this description, the structure of biomolecules is rather complex and investigations of their properties by analytical methods are hard to perform. What can we do in that case? To answer the question, let us recall analogous situations in physics. When physicists study complex systems consisting of many interacting particles, they act in the following way: they construct a simplified model of the real system, which could be accurately treated by mathematical methods. Such an approach gives them a confidence that the results obtained are the consequence of the assumptions of the model and are not distorted by mistakes arising from an approximate calculation. This is the main idea of the model approach in physics.

Following this logic we naturally come to the problem of constructing the models of biomolecules. We shall assume that the models have not to take into account all the details of biomolecular structure, because such a statement of the problem would be too complex. The models should only reproduce the most important (or dominant) properties of biomolecules. What are these properties?

1. Reading paragraphs 1 and 2 one can notice at least two general characteristics of the biomolecular structures. The first is that biomolecules of the two considered classes consists of long chains of atoms. The second is

that these chains have nearly regular structure, that is biomolecules have a "skeleton" with accurately repeating pattern of atoms along the chains. Due to these properties biomolecules are similar to a certain extent to one-dimensional periodical structures which are known in physics as quasi - one-dimensional crystals. This is why Charls Bunn gave them a poetical name: "crystals of chains of life" [15]. Taking into account these observations we can consider one-dimensional crystal as he simplest physical analog (or model) of biomolecules.

2 However in some aspects biomolecules are more similar to polymers than to crystals, because in addition to the properties mentioned above, biomolecules are not rigid systems, but flexible ones. This property is reflected in the second name of biomolecules - biopolymers. So, if we want to construct a more accurate model, we must take into account the flexible nature of biomolecules, that is their ability to bend, to twist, to form superstructures and so on.

3. Besides a "skeleton" with a regular alteration of atoms or atomic groups, biomolecules have elements of irregular structure. So, if we want to improve the model, we must take into account the irregularity of the side chains in the case of proteins and that of the base sequence in the case of DNA. We can consider these irregularities as a small disturbance of the regular pattern of the "skeleton" and use perturbation theory for mathematical treatment

Now, even from these simple argument we can conclude that there is the possibility to construct a hierarchy of models describing biomolecules with different degrees of accuracy. At the beginning of the book the reader will meet the simplest physical analogs of biomolecules - one-dimensional crystals (or chains), then these models will little by little become more complicated and accurate because more and more details of the structure and interactions will be included.

PART 2. METHODS OF QUANTUM STATISTICAL PHYSICS.

As follows from the material of the previous chapter, biomolecules are systems consisting of hundreds, thousands and more interacting particles (atoms), organized in some special way in space. So, it is clear that any attempt to investigate biomolecules by the methods of theoretical physics will be associated with the necessity to use the methods of so-called "theory of systems of many particles " or "statistical physics".

In this chapter we present a brief review of the basic methods of one of the important parts of the theory, namely, the quantum statistics. The description of the methods does not pretend to completeness, it includes only minimal information necessary to understand applications. The reader can find a more detailed description of methods of quantum statistical theory in the books of Bogolubov [16], Tyablikov [17] and in the review of Zubarev [18].

§4. Hamiltonian functions.

In the previous chapter a brief description of possible models of biomolecules has been given. Now, we shall try to translate it into mathematical language. The translation can be made in many different ways. Two of them are the most known and widely used in classical physics. These are the formalism of Lagrange and that of Hamilton [19]. These formalisms differ from each other by the way in which the state of a dynamical system is determined and the equation of evolution is derived.

In the first formalism the state of a dynamical system consisting of N particles is completely and uniquely determined by 3N generalized coordinates $q_1, q_2, q_3 \dots q_N$ and 3N velocities $\dot{q}_1, \dot{q}_2, \dot{q}_3 \dots \dot{q}_N$. For short, 3N coordinates and 3N velocities are usually denoted by symbols q and $\dot{q}$, respectively. The dynamics (or evolution) of the system is determined then by the Langrange function (or lagrangian)

$$L = L(q, \dot{q}, t), \tag{4-1}$$

which is given in the space of variables q, $\dot{q}$ and t. With the help of the function L one can construct dynamical equations according to the rule

$$\frac{d}{dt}\frac{\partial L}{\partial \dot{q}_j} - \frac{\partial L}{\partial q_j} = 0, \qquad j = 1, 2, \dots 3N. \tag{4-2}$$

Equations (4-2) are known as Lagrange equations. They can be deduced from a more general principle of classical physics known as variation principle of Hamilton. According to this principle, the motion of any dynamic system in the time interval from t_1 to t_2 is determined by the requirement that the variation of the integral $I = \int_{t_2}^{t_1} L(q, \dot{q}, t)dt$ should be equal to zero

$$\delta I = 0. \tag{4-3}$$

In the second formalism the state of a system of N particles is determined by 3N generalized coordinates $q_1, q_2, q_3 \ldots q_N$ and 3N conjugate impulses $p_1, p_2, p_3 \ldots p_N$. For short we shall denote them by symbols q and p, correspondingly. The dynamics of the system is determined then by the Hamilton function (or hamiltonian)

$$\mathcal{H}(p, q, t), \tag{4-4}$$

given in the space of variables q, p and t. With the help of the function $\mathcal{H}$ one can construct the dynamical equations according to the rule

$$\dot{q}_j = \frac{\partial \mathcal{H}}{\partial p_j}, \quad -\dot{p}_j = \frac{\partial \mathcal{H}}{\partial q_j}, \quad j = 1, 2, \ldots 3N. \tag{4-5}$$

Equations (4-5) are known as canonical equations of Hamilton. As in the first case, the dynamical equations can be deduced from the general variation principle (4-3).

To go from one way of description to the other one, for example from the formalism of Lagrange to the formalism of Hamilton, it is sufficient to pass from variables q, $\dot{q}$, t to variables q, p, t and to construct the function $\mathcal{H}$ instead of the function L, the new function and new variables being related to old ones by the following formulas

$$p_j = \frac{\partial L}{\partial \dot{q}_j},$$

$$\mathcal{H} = \sum_j \dot{q}_j p_j - L(q, \dot{q}, t), \quad j = 1, 2, \ldots 3N. \tag{4-6}$$

In this and in the next chapters we shall give preference to the formalism of Hamilton. Using this single and universal mathematical language we shall describe the methods of quantum statistical theory and their applications. Thus we shall put an additional sense into the words "constructing the model". This will mean now not only the choice of a more or less suitable physical model of biomolecule, but also the construction of the model hamiltonian $\mathcal{H}$. In every particular case, the form of the function $\mathcal{H}$ will be completely determined by the structure of, and interactions in, the system. For example, in the case of a system of N identical particles moving in three-dimensional space under the action of the potential field V(q), the model hamiltonian has the following simple form

$$\mathcal{H} = \sum_{j=1}^{3N} \frac{p_j^2}{2m} + V(q). \qquad (4\text{-}7)$$

where m is the mass of particles.

One can ask: why did we choose the formalism of Hamilton? The main reason is that this formalism can be successfully used in both classical and quantum physics. However we must note that in quantum physics the formalism of Hamilton is used in some modified form. Let us consider the basic elements of this formalism in the case of quantum systems. Here, in contrast to the classical case, the state of a quantum system at time t is described by a complex wave function $\Psi(q,t)$, the squared module of which ($|\Psi(q,t)|^2$) is interpreted as the probability of the system to be in this state.

It is obvious that $\int |\Psi(q,t)|^2 dq=1$. The dynamics of a quantum system is determined by an operator $\hat{\mathcal{H}}$ which operates in the space of the wave functions Ψ. (the hat ^ points out that $\hat{\mathcal{H}}$ is not an ordinary function, but an operator). With the help of the hamiltonian operator $\hat{\mathcal{H}}$ one can obtain the dynamical equation describing the evolution of the quantum system

$$i\hbar \frac{\partial \Psi}{\partial t} = \hat{\mathcal{H}} \Psi , \tag{4-8}$$

which is known as a "wave equation of Schrödinger". Here is $\hbar$ the Planck constant divided by 2π.

To illustrate the material described above, let us return again to the example of a dynamical system consisting of N identical particles moving under the action of the potential field V(q). The Hamilton operator for this system has the following form

$$\hat{\mathcal{H}} = - \frac{\hbar^2}{2m} \sum_{j=1}^{3N} \frac{\partial^2}{\partial q_j^2} + V(q) . \tag{4-9}$$

Note that in this case the hamiltonian $\hat{\mathcal{H}}$ is not explicitly dependent on time; therefore the evolution equation (4-8) has a solution with separable variables

$$\Psi(q,t) = \Phi(t)\,\varphi(q) . \tag{4-10}$$

Substituting (4-10) into (4-8), we find

$$i\hbar \frac{d\Phi/dt}{\Phi} = \frac{\hat{\mathcal{H}}.\varphi}{\varphi} = E , \tag{4-11}$$

where E is a constant. The solution of the first of the two equations (4-11) has the form

$$\Phi(t) = \exp(iEt/\hbar). \qquad (4\text{-}12)$$

It remains only to find the solution of the second equation. Thus, the dynamical problem (4-8) reduces now to a simpler one

$$\hat{\mathcal{H}}\varphi = E\varphi. \qquad (4\text{-}13)$$

which can be formulated as the problem of finding eigenfunctions and eigenvalues of the operator of Hamilton $\hat{\mathcal{H}}$. Equation (4-13) is known in quantum physics as a "stationary Schrödinger equation".

§5. Method of secondary quantization.

In the previous paragraph the two main ways of mathematical description of the state and evolution of dynamical systems were given. However, if a system consists of identical particles, its state and evolution can be described in another way, which is usually named the method of secondary quantization. Indeed, if a system consists of N identical particles, each being only in one of several discrete states, and if we denote the number of particles being in state 1 by N_1, the number of particles being in state 2 by N_2, and so on, then it is easy to notice that the set of numbers

$$N_1, N_2, \ldots N_l, \ldots \tag{5-1}$$

will completely and uniquely describe the state of the whole system. The numbers $N_1, N_2, \ldots N_l \ldots$ are usually named occupation numbers. According to the standard way of description of the dynamical system states accepted in quantum physics, we can determine the "wave" functions in the space of occupation numbers $N_1, N_2, \ldots N_l, \ldots$ Usually, these functions are named "the wave functions of secondary quantization", and denoted by the symbol

$$|\ldots N_l \ldots\rangle \tag{5-2}$$

The squared module of (5-2) is interpreted as the probability that N_1 particles of the system are in state 1, N_2 particles are in state 2 and so on. Evolution (or dynamics) will be determined in this case by the equation

$$i\hbar \frac{\partial}{\partial t} |\ldots N_l \ldots\rangle = \hat{\mathcal{H}} |\ldots N_l \ldots\rangle. \tag{5-3}$$

where $\hat{\mathcal{H}}$ is the operator of Hamilton written in terms of the secondary quantum representation. Now it remains only to discuss the method by which the operator of Hamilton or some other operator can be written in terms of the secondary quantum representation.

There is an algorithm for rewriting the operators which act in the space of ordinary wave functions Ψ in terms of secondary quantization. This algorithm is well described by Bogolubov [16]. However, in [16] as well as in the majority of other books only two cases are considered: when the wave function Ψ is symmetric and when it is antisymmetric relative to the transposition of any two particles. In the first case the operator of Hamilton can be written in terms of Bose-operators $\hat{b}^{+}, \hat{b}$, and about these systems investigators say that they obey the Bose-statistics. In the second case the operator of Hamilton can be written in terms of Fermi-operators $\hat{a}^{+}, \hat{a}$, and about these systems it is usually said that they obey the Fermi-statistics. Systems of photons or phonons are examples of the systems of the first type, a system of electrons is an example of the second type of system.

However, the variety of possible symmetries is not confined by complete symmetry and complete antisymmetry. The existence of the so-called Pauli-statistics confirms this statement. In this case usual quantum operators acting in the space of wave functions Ψ, can be expressed in terms of Pauli-operators $\hat{p}^{+}, \hat{p}$. For illustration, let us consider a rather general example described in the monograph of Agranovich [20].

Let us consider a crystal with immovable (i.e. with hardly

fixed in the equilibrium positions) molecules. Assume that the elementary crystal cell contains σ molecules and introduce the index α to number the molecules inside the cell ($\alpha=1,2,\ldots\sigma$). Denote also the lattice translation vector by symbol $\vec{n}$

$$\vec{n} = n_1\vec{a}_1 + n_2\vec{a}_2 + n_3\vec{a}_3 , \tag{5-4}$$

where n_1, n_2, n_3 are the integral numbers, $\vec{a}_1, \vec{a}_2, \vec{a}_3$ are the unit basis vectors. Then the operator of Hamilton (hamiltonian) of a crystal can be easy written in usual (Schrödinger's) representation

$$\hat{\mathcal{H}} = \sum_{\vec{n},\alpha} \hat{H}_{\vec{n},\alpha} + \frac{1}{2}\sum'_{\vec{n},\alpha;\vec{m},\beta} \hat{V}_{\vec{n},\alpha;\vec{m},\beta} , \tag{5-5}$$

where $\hat{H}_{n,\alpha}$ is the hamiltonian of the n,α-th isolated molecule, $\hat{V}_{\vec{n},\alpha;\vec{m},\beta}$ is the operator of interaction of the $\vec{n},\alpha$-th and $\vec{m},\beta$-th molecules. The apostrophe which is near one of the two symbols of summation in formula (5-5), means that in this sum the terms $\hat{V}_{\vec{n},\alpha;\vec{m},\beta}$ with $\vec{n},\alpha=\vec{m},\beta$ are absent. For simplicity, we shall use in the following more convenient designations: $n=\{\vec{n},\alpha\}$ and $m=\{\vec{m},\beta\}$.

Let us assume that the solutions of the stationary Schrödinger's equation

$$\hat{H}_n \varphi_{n,f} = E_f \varphi_{n,f} \tag{5-6}$$

for an isolated free molecule are known. Here $\varphi_{n,f}$ and E_f are the eigenfunctions and eigenvalues of the operator $\hat{H}_n$, respectively. A scheme illustrating a possible arrangement of energy levels in this case is shown in Fig. 12.

Let us introduce then the occupation numbers $N_{n,f}$ according to the rule

$$N_{n,f} = \begin{cases} 1, & \text{if the n-th molecule is in the state } \varphi_{n,f} \\ 0, & \text{if the n-th molecule is not in the state } \varphi_{n,f} . \end{cases} \qquad (5\text{-}7)$$

As every molecule can be only in one of the stationary states, the occupation numbers must satisfy the requirement

$$\sum_f N_{n,f} = 1 . \qquad (5\text{-}8)$$

If we take into account that the total number of molecules in the crystal is equal to σN, it becomes clear that the occupation numbers $N_{n,f}$ must also satisfy the requirement

$$\sum_{n,f} N_{n,f} = \sigma N . \qquad (5\text{-}9)$$

We shall characterize different states of the crystal by the wave function of secondary quantization

$$| \ldots N_{n,f} \ldots \rangle . \qquad (5\text{-}10)$$

Let us construct the operators acting on the wave functions (5-10). We shall begin with the simplest operator, namely, the operator of number of states $\hat{N}_{n,f}$. In the representation of the occupation numbers this operator is diagonal, and its eigenvalues are the occupation numbers $N_{n,f}$ themselves

$$\hat{N}_{n,f} | \ldots N_{n,f} \ldots \rangle = N_{n,f} | \ldots N_{n,f} \ldots \rangle . \qquad (5\text{-}11)$$

Let us express the Hermitian operator $\hat{N}_{n,f}$ through the product of two operators which are not Hermitian

$$\hat{N}_{n,f} = \hat{P}^{+}_{n,f}\hat{P}_{n,f} . \qquad (5\text{-}12)$$

To avoid the breaking of the equality (5-11) it is enough to require that the operation of $\hat{p}^{+}_{n,f}$ and $\hat{p}_{n,f}$ on the wave function of secondary quantization (5-10) is determined by the following rules.

$$\hat{P}^{+}_{n,f}|\ldots N_{n,f}\ldots\rangle = (1-N_{n,f})|\ldots(1+N_{n,f})\ldots\rangle, \qquad (5\text{-}13)$$

$$\hat{P}_{n,f}|\ldots N_{n,f}\ldots\rangle = N_{n,f}\,|\ldots(1-N_{n,f})\ldots\rangle. \qquad (5\text{-}14)$$

Equations (5-13) and (5-14) permit us to call the operators $\hat{p}^{+}_{n,f}$, $\hat{p}_{n,f}$ the operators of creation of the state $\varphi_{n,f}$ in the crystal (this case corresponds to the transition of the n-th molecule from the ground state to the excited state with energy E_f), and of annihilation of the same state, respectively.

As it appears also from equations (5-13)-(5-14), the operators $\hat{p}^{+}_{n,f}$, $\hat{p}_{n,f}$ with the identical indexes n, f should satisfy the following commutation rules,

$$\hat{P}^{+}_{n,f}\hat{P}_{n,f} + \hat{P}_{n,f}\hat{P}^{+}_{n,f} \equiv [\hat{P}^{+}_{n,f}, \hat{P}_{n,f}]_{+} = 1 , \quad \hat{P}_{n,f}\hat{P}_{n,f} = 0. \qquad (5\text{-}15)$$

At the same time, the operators $\hat{p}^{+}_{n,f}$, $\hat{p}_{m,g}$ with different indexes ($n \neq m$ or $f \neq g$), which hence act on different variables of the wave functions $|\ldots N_{n,f} \ldots N_{m,g} \ldots\rangle$, are commutative, i.e.

$$\hat{p}_{n,f}\hat{p}^{+}_{m,g} - \hat{p}^{+}_{m,g}\hat{p}_{n,f} \equiv [\hat{p}_{n,f}, \hat{p}^{+}_{m,g}] = 0, \quad [\hat{p}_{n,f}, \hat{p}_{m,g}] = 0, \qquad (5\text{-}16)$$

if $n \neq m$ or $f \neq g$.

The reader who knows the commutation rules for operators of Bose and Fermi, must notice that the operators $\hat{p}^{+}_{n,f}$, $\hat{p}_{m,g}$ referring to the same excited state ($f=g$) and the same molecule ($n=m$), satisfy Fermi's commutation rules (see eq. (5-15)). In the case when these operators refer to different molecules ($n \neq m$) or to different states ($f \neq g$), they satisfy Bose's commutation rules (see eq. (5-16)). This mixed statistics is named Pauli-statistics, and the operators $\hat{p}^{+}_{n,f}$ $\hat{p}_{n,f}$ are named Pauli-operators.

To formulate the rules of transformation of the operators written in terms of Schrödinger's representation, to those written in terms of secondary quantum representation, it remains now to introduce only the so-called operator functions $\hat{\Psi}^{+}$, $\hat{\Psi}$

$$\hat{\Psi}(q_n) = \sum_{n,f} \hat{p}_{n,f}\varphi_{n,f}(q_n), \qquad (5\text{-}17)$$

$$\hat{\Psi}^{+}(q_n) = \sum_{n,f} \hat{p}^{+}_{n,f}\varphi^{*}_{n,f}(q_n).$$

Then the transformation to the secondary quantum representation will be determined by the following equations

$$(1) \quad \hat{H}_n = \int \hat{\Psi}^{+}(q_n)\hat{H}_n\hat{\Psi}(q_n)\,dq_n = \qquad (5\text{-}18)$$

$$= \sum_{f,g} \hat{p}^{+}_{n,f}\langle\varphi_{n,f}|\hat{H}_n|\varphi_{n,g}\rangle\hat{p}_{n,g} = \sum_{f} E_f\hat{p}^{+}_{n,f}\hat{p}_{n,f},$$

$$(2)\quad \hat{V}_{n,m} = \iint \hat{\Psi}^{+}(q_n)\hat{\Psi}^{+}(q_m)\hat{V}_{n,m}\hat{\Psi}(q_m)\hat{\Psi}(q_n)\,dq_n\,dq_m = \tag{5-19}$$

$$= \sum_{f,g,f',g'}' \langle \varphi_{n,f'}\varphi_{m,g'}|\hat{V}_{n,m}|\varphi_{n,f}\varphi_{m,g}\rangle \hat{P}^{+}_{n,f'}\hat{P}^{+}_{m,g'}\hat{P}_{m,g}\hat{P}_{n,f},$$

and so on.

Using the rules (5-18)-(5-19) we can easily write hamiltonian (5-5) in the secondary quantum representation

$$\hat{\mathcal{H}} = \sum_{n,f} E_f \hat{P}^{+}_{n,f}\hat{P}_{n,f} + \frac{1}{2}\sum_{n,m}' \sum_{f,f',g,g'} \langle f'g'|\hat{V}_{n,m}|g,f\rangle \hat{P}^{+}_{n,f'}\hat{P}^{+}_{m,g'}\hat{P}_{m,g}\hat{P}_{n,f}, \tag{5-20}$$

where f, f', g, g' are the quantum numbers characterizing the stationary states of molecules, and $\langle f'g'|\hat{V}_{n,m}|g,f\rangle$ is a shorter notation for $\langle \varphi_{n,f'}\varphi_{m,g'}|\hat{V}_{n,m}|\varphi_{n,f}\varphi_{m,g}\rangle$.

Hamiltonian (5-20) can be simplified in the case when not all, but only two possible states of the isolated molecule are taken into account: the ground state (0) and the first excited one (1). In this case only the terms with the indexes f, f', g and g' equal to 0 or 1 remain in the sums of hamiltonian (5-20)

$$\hat{\mathcal{H}} = \sum_{n} \Delta_1 \hat{P}^{+}_{n,1}\hat{P}_{n,1} + \sum_{n,m}' \mathcal{M}^{1}_{n,m} \hat{P}^{+}_{n,0}\hat{P}^{+}_{m,1}\hat{P}_{m,0}\hat{P}_{n,1} +$$

$$+ \frac{1}{2}\sum_{n,m}' \mathcal{M}'^{1}_{n,m} \{\hat{P}^{+}_{n,0}\hat{P}^{+}_{m,0}\hat{P}_{m,1}\hat{P}_{n,1} + \hat{P}^{+}_{n,1}\hat{P}^{+}_{m,1}\hat{P}_{m,0}\hat{P}_{n,0}\} + \tag{5-21}$$

$$+ \frac{1}{2}\sum_{n,m}' \mathcal{L}^{1}_{n,m} \hat{P}^{+}_{n,1}\hat{P}^{+}_{m,1}\hat{P}_{m,1}\hat{P}_{n,1},$$

where $\Delta_1 = E_1 - E_0 + \sum_{m}' \mathcal{D}_{n,m}$, $\mathcal{D}_{n,m} = \langle 01|\hat{V}_{n,m}|01\rangle - \langle 00|\hat{V}_{n,m}|00\rangle$,

$\mathcal{M}^{1}_{n,m} = \langle 01|\hat{V}_{n,m}|10\rangle$, $\mathcal{M}'^{1}_{n,m} = \langle 00|\hat{V}_{n,m}|11\rangle = \langle 11|\hat{V}_{n,m}|00\rangle$,

$\mathcal{L}^{1}_{n,m} = \langle 00|\hat{V}_{n,m}|00\rangle + \langle 11|\hat{V}_{n,m}|11\rangle - 2\langle 10|\hat{V}_{n,m}|10\rangle.$

Let us introduce new operators

$$\hat{P}_{n,1} = \hat{p}^{+}_{n,0}\,\hat{p}_{n,1}\,,$$

$$\hat{P}^{+}_{n,1} = \hat{p}^{+}_{n,1}\,\hat{p}_{n,0}\,. \tag{5-22}$$

It is easy to check that the operators $\hat{P}^{+}$, $\hat{P}$ are Pauli-operators; that is, they satisfy commutation rules (5-15)-(5-16). In terms of the operators $\hat{P}^{+}$, $\hat{P}$ hamiltonian (5-21) takes a substantially simpler form

$$\hat{\mathcal{H}} = \sum_n \Delta\,\hat{P}^{+}_n\hat{P}_n + \sum_n{}' \mathcal{M}_{n,m}\,\hat{P}^{+}_m\hat{P}_n + \frac{1}{2}\sum_{n,m}{}' \mathcal{M}'_{n,m}\,(\hat{P}^{+}_m\hat{P}^{+}_n + \hat{P}_m\hat{P}_n) + \frac{1}{2}\sum_{n,m}{}' \mathcal{I}_{n,m}\,\hat{P}^{+}_n\hat{P}_n\hat{P}^{+}_m\hat{P}_m\,. \tag{5-23}$$

For simplicity, we omitted here index 1 in all coefficients and operators.

§6. Transformation from Pauli-operators to Fermi- and Bose- ones.

In contrast to Pauli-operators, Fermi-operators satisfy commutation rules of the form

$$[\hat{a}_l, \hat{a}^+_{l'}]_+ = \delta_{l,l'}, \quad [\hat{a}_l, \hat{a}_{l'}]_+ = 0, \tag{6-1}$$

$N_l = 0, 1;$

and Bose-operators satisfy commutation rules of the form

$$[\hat{b}_l, \hat{b}^+_{l'}]_- = \delta_{l,l'}, \quad [\hat{b}_l, \hat{b}_{l'}]_- = 0, \tag{6-2}$$

$N_l = 0, 1, 2, \ldots$

Here, one or several indexes which numerate excited states of the system, are designated by l, l'; $\delta_{l,l'}$ is Kroneker symbol: $\delta_{l,l'} = 0$, if $l \neq l'$, $\delta_{l,l'} = 1$, if $l = l'$.

In spite of the difference in commutation rules (6-1), (6-2) and (5-15), (5-16), the exact and approximate relations between Pauli-, Bose- and Fermi-operators can be found. Let us consider the relation between Pauli- and Bose-operators. For this purpose, let us rewrite the operational equality

$$\hat{P}_l \hat{P}^+_l + \hat{P}^+_l \hat{P}_l = 1 \tag{6-3}$$

in the form

$$\hat{P}_l \hat{P}^+_l - \hat{P}^+_l \hat{P}_l = 1 - 2\hat{N}_l. \tag{6-4}$$

In the case when only a small part of crystal molecules are excited, i.e. when the condition for the average value of the occupation number N_l

$$\langle N_\ell \rangle \ll 1 \tag{6-5}$$

is fulfilled, we can assume that operators $\hat{P}^+_\ell, \hat{P}_{\ell'}$ satisfy with a good accuracy the Bose's commutation rules

$$\hat{P}_\ell \cong \hat{b}_\ell, \quad \hat{P}^+_\ell \cong \hat{b}^+_\ell,$$
$$[\hat{b}_\ell, \hat{b}^+_{\ell'}]_- = \delta_{\ell,\ell'}. \tag{6-6}$$

This approximation is widely used in studying the excitons in crystals [20,21]. However, the reader should remember that the substitution of Pauli-operators by Bose-ones is a rather approximate procedure and it has one essential defect; namely, the occupation numbers N_ℓ for paulions take only two values: 0 and 1 (see (5-7)), while the occupation numbers N_ℓ for bosons take any positive values: 0, 1, 2... (see (6-2)). Therefore, when substituting the operators $\hat{P}^+_\ell, \hat{P}_\ell$ by $\hat{b}^+_\ell, \hat{b}_\ell$ uncontrolled mistakes could be inserted if the number of bosons is more than 1.

The method of transformation from Pauli-operators to Bose-ones described above can be improved. For this purpose, let us assume that the number of paulions is equal to 0 or 1 while the number of bosons is arbitrary. Then we can rewrite Pauli-operators in the form

$$\hat{P}_\ell = \left(\sum_{\nu=0}^{\infty} \alpha_\nu \hat{b}^{+\nu}_\ell \hat{b}^{\nu}_\ell\right)^{1/2} \hat{b}_\ell,$$
$$\hat{P}^+_\ell = \hat{b}^+_\ell \left(\sum_{\nu=0}^{\infty} \alpha_\nu \hat{b}^{+\nu}_\ell \hat{b}^{\nu}_\ell\right)^{1/2}, \tag{6-7}$$

where are real coefficients. Let us assume also that operators $\hat{P}^+_\ell, \hat{P}_\ell$ satisfy the condition

$$\hat{P}_\ell \hat{P}_\ell^+ + \hat{P}_\ell^+ \hat{P}_\ell = 1. \tag{6-8}$$

Substituting (6-7) into (6-8) and taking into account the equality

$$\hat{b}_\ell^{+\nu} \hat{b}_\ell^{\nu+1} = (\hat{N}_\ell - \nu)\, \hat{b}_\ell^{+\nu} \hat{b}_\ell^{\nu}, \tag{6-9}$$

we obtain the following formula for the coefficients

$$\alpha_\nu = (-2)^\nu / (1+\nu)! \tag{6-10}$$

Thus, the accurate transformation from Pauli-operators to Bose-ones will have the form

$$\hat{P}_\ell = \Big[\sum_{\nu=0}^{\infty} \frac{(-2)^\nu}{(1+\nu)!} \hat{b}_\ell^{+\nu} \hat{b}_\ell^{\nu} \Big]^{1/2} \hat{b}_\ell ,$$

$$\hat{P}_\ell^+ = \hat{b}_\ell^+ \Big[\sum_{\nu=0}^{\infty} \frac{(-2)^\nu}{(1+\nu)!} \hat{b}_\ell^{+\nu} \hat{b}_\ell^{\nu} \Big]^{1/2}, \tag{6-11}$$

If we confine ourselves to the first members in expansions (6-11) we obtain the approximate formulas

$$\hat{P}_\ell \cong \hat{b}_\ell , \quad \hat{P}_\ell^+ \cong \hat{b}_\ell^+ , \tag{6-12}$$

which are equivalent to approximation (6-6), considered above. If we confine ourselves to two members (ν=0, 1) in the expansions (6-11), we obtain the formulas

$$\hat{P}_\ell \cong \sqrt{1-\hat{N}_\ell}\;\hat{b}_\ell, \qquad \hat{P}^{+}_\ell \cong \hat{b}^{+}_\ell \sqrt{1-\hat{N}_\ell}, \tag{6-13}$$

which are equivalent to the approximation known in physics as Holstein-Primakoff approximation [22].

Now let us pass to the next question concerning the transformation of Pauli-operators to Fermi-ones. As shown by Chesnut and Suna [23], these relations can be easily found in one-dimensional case. For this purpose it is sufficient to compare the action of Pauli-operators and Fermi-operators on the wave functions of the secondary quantization $|\ldots N_\ell \ldots\rangle$. Corresponding equations for Pauli-operators are given in 5 (see formulas (5-13) and (5-14)). Analogous equations for Fermi-operators have the form [24]

$$\hat{a}_\ell |\ldots N_\ell \ldots\rangle = (-1)^{\sum_{\nu=1}^{\ell-1} N_\nu} N_\ell |\ldots (N_\ell - 1)\ldots\rangle, \tag{6-14}$$

$$\hat{a}^{+}_\ell |\ldots N_\ell \ldots\rangle = (-1)^{\sum_{\nu=1}^{\ell-1} N_\nu} (1-N_\ell)|\ldots (N_\ell + 1)\ldots\rangle. \tag{6-15}$$

Comparing equations (6-14), (6-15) with equations (5-13), (5-14) one can obtain the following formulas

$$\hat{a}_\ell = \hat{P}_\ell \hat{\epsilon}_\ell, \qquad \hat{a}^{+}_\ell = \hat{P}^{+}_\ell \epsilon_\ell, \tag{6-16}$$

which give us the rule of transformation of Pauli-operators $\hat{P}^{+}_\ell, \hat{P}_\ell$ to Fermi-operators $\hat{a}^{+}_\ell, \hat{a}_\ell$. Here, the operator $\hat{\epsilon}_1$ is determined by the following expression

$$\hat{\epsilon}_\ell |\ldots N_\ell \ldots\rangle = \prod_{\nu=1}^{\ell-1} (-1)^{N_\nu} |\ldots N_\ell \ldots\rangle. \tag{6-17}$$

It is easy to check now that Pauli-operators $\hat{P}^+_\ell, \hat{P}_\ell$ commute with the operator $\hat{\epsilon}_{\ell'}$ if $\ell \geq \ell'$, and anticommutate with the operator $\hat{\epsilon}_{\ell'}$ if $\ell < \ell'$.

$$[\hat{P}_\ell, \hat{\epsilon}_{\ell'}]_- = 0, \quad [\hat{P}^+_\ell, \hat{\epsilon}_{\ell'}]_- = 0, \qquad if \;\; \ell \geq \ell', \tag{6-18}$$

$$[\hat{P}_\ell, \hat{\epsilon}_{\ell'}]_+ = 0, \; [\hat{P}^+_\ell, \hat{\epsilon}_{\ell'}]_+ = 0 \;, \quad if \;\; \ell < \ell'. \tag{6-19}$$

Using formulas (6-16) one can show also that sometimes the system of paulions (5-23) can exhibit Fermi-properties. In § 10-15 the reader will meet some examples.

7. Heisenberg representation.

Two equivalent ways of describing the behavior of dynamical systems as a function of time are possible in quantum mechanical theory. According to the first one, it is assumed that the operators given in the space of wave functions of the system do not depend explicitly on time t, and all time dependence is focused in the wave functions. If we take into account the independence of hamiltonian $\hat{\mathcal{H}}$ upon the time t, we can easily obtain from eq. (4-8) the relation of the wave function $\Psi(q,0)$ which determines the state of the system at a moment t=0, to the wave function $\Psi(q,t)$ describing the state of the system at a moment t (t=0),*

$$\Psi(q,t) = \Psi(q,0)\,e^{-i\hat{\mathcal{H}}t} \qquad (7\text{-}1)$$

This way of describing is named Schrödinger representation. However, it is more convenient to use the other way, in which the wave functions Ψ are assumed to be independent of time, and all time dependence is focused in operators. This way of describing is named Heisenberg representation.

It is not difficult to find the relation between the operators written in Schrödinger representation and those written in Heisenberg one. For this purpose we can make use of the natural requirement that the quantum mechanical averages of the observable physical values (i.e. of operators) should be independent of the type of representation. Let us assume that $\hat{A}(0)$ is an operator. In Schrödinger representation this operator does not depend

---

* For simplicity, we assume that $\hbar=1$.

explicitly on time, and its average is determined by formula

$$\langle \hat{A}(0) \rangle_S = \int \Psi^*(q,t) \hat{A}(0) \Psi(q,t) dq . \tag{7-2}$$

Inserting (7-1) into (7-2) we obtain

$$\langle \hat{A}(0) \rangle_S = \int \Psi^*(q,0)\, e^{i\hat{\mathcal{H}}t} \hat{A}(0) e^{-i\hat{\mathcal{H}}t} \Psi(q,0) dq . \tag{7-3}$$

In Heisenberg representation the operator $\hat{A}(t)$ depends on time t, and its average is determined by formula

$$\langle \hat{A}(t) \rangle_H = \int \Psi^*(q,0) \hat{A}(t) \Psi(q,0) dq . \tag{7-4}$$

Using the requirement

$$\langle \hat{A}(t) \rangle_H = \langle \hat{A}(0) \rangle_S \tag{7-5}$$

we find the relation between the operators written in Schrödinger and Heisenberg representation

$$\hat{A}(t) = e^{i\hat{\mathcal{H}}t} \hat{A}(0) e^{-i\hat{\mathcal{H}}t} . \tag{7-6}$$

Now it remains only to obtain the dynamical equations (or equations of evolution) for the operators in Heisenberg representation. (The dynamical equation for the wave function has been discussed already in §4). For this purpose, let us consider again the quantum mechanical averages and write the change (in time) of the average of the operator $\hat{A}$ written in Schrödinger representation

$$\frac{d}{dt}\langle \hat{A}(0)\rangle_S = \frac{d}{dt}\int \Psi^*(q,t)\hat{A}(0)\Psi(q,t)dq \,. \tag{7-7}$$

Differentiating the right hand side of eq. (7-7) and taking into account eq. (7-1), we obtain

$$\frac{d}{dt}\langle \hat{A}(0)\rangle_S = \int \Psi^*(q,0)\frac{d}{dt}\left(e^{i\hat{\mathcal{H}}t}\hat{A}(0)e^{-i\hat{\mathcal{H}}t}\right)\Psi(q,0) =$$

$$= i\int \Psi^*(q,0)e^{i\hat{\mathcal{H}}t}\left(\hat{\mathcal{H}}\hat{A}(0) - \hat{A}(0)\hat{\mathcal{H}}\right)e^{-i\hat{\mathcal{H}}t}\Psi(q,0). \tag{7-8}$$

Then, using the natural equality: $\frac{d}{dt}\langle\hat{A}(0)\rangle_S = \frac{d}{dt}\langle\hat{A}(t)\rangle_H$, we find the dynamical equation for the operator $\hat{A}$ in Heisenberg representation

$$i\frac{d}{dt}\hat{A}(t) = [\hat{A}(t), \hat{\mathcal{H}}] \,. \tag{7-9}$$

By the same method one can consider also two ways of describing of the dynamics of quantum systems in terms of secondary quantization. It is easy to make sure that eq. (7-9) has the same form for the operators written in terms of the secondary quantum representation.

Operator equation (7-9) has an analog in classical physics. To demonstrate this statement, let us consider a system consisting of N particles. Let us assume also that the state of the system is completely determined by 3N generalized coordinates $q_1, q_2, \ldots q_{3N}$ and 3N impulses $p_1, p_2, \ldots p_{3N}$. Let A be a physical quantity. In the general case, A is a function of 3N coordinates $q_1, q_2, \dots q_{3N}$, 3N impulses $p_1, p_2, \dots p_{3N}$ and time t. The total time differential of A in time has then the following form

$$\frac{dA}{dt} = \frac{\partial A}{\partial t} + \sum_{j=1}^{3N} \left(\frac{\partial A}{\partial q_j} \dot{q}_j + \frac{\partial A}{\partial p_j} \dot{p}_j \right). \qquad (7\text{-}10)$$

Inserting eqs. (4-5) into (7-10), we find

$$\frac{dA}{dt} = \frac{\partial A}{\partial t} + \{\mathcal{H}, A\}, \qquad (7\text{-}11)$$

where {...} are Poisson brackets. In the case when A does not depend explicitly on time, we obtain

$$\frac{dA}{dt} = \{\mathcal{H}, A\}, \qquad (7\text{-}12)$$

which is known as the "classical" analog of eq. (7-9).

§8. Time correlation functions.

When studying a system consisting of a large number of interacting particles, the so-called time correlation functions are very important. Usually they are determined as a product of operators written in Heisenberg representation, which is averaged over a statistical ensemble

$$F_{\hat{A}\hat{B}}(t,t') = \langle \hat{A}(t)\hat{B}(t')\rangle = \frac{Sp\{\hat{A}(t)\hat{B}(t')e^{-\hat{\mathcal{H}}/k_BT}\}}{Sp\{e^{-\hat{\mathcal{H}}/k_BT}\}} \quad . \qquad (8-1)$$

Here $\hat{A}(t)$, $\hat{B}(t')$ are the operators in Heisenberg representation (in the general case they are the products of operators of creation and annihilation of particles), k_B is the Boltzmann constant, T is temperature. Let $\hat{\mathcal{H}}$ include also the term $-\mu N$, where μ is the chemical potential, and N is the operator of the total number of particles. The symbol of spur (Sp) denotes summation over all diagonal elements of the matrix of the corresponding operator, the summation being taken both over the number of particles and over all possible states with a given number of particles. The symbol <...> was used in the previous paragraphs to denote the quantum mechanical averages. However, here (and in further paragraphs) this symbol denotes statistical averaging over the grand canonical ensemble.

It can be shown that the correlation functions (8-1) depend only on the difference of time arguments: $F_{\hat{A}\hat{B}}(t,t')=F_{\hat{A}\hat{B}}(t-t')$, and in the case when the arguments coincide (t=t') the correlation functions transform to the usual statistical averages and they do not depend on time: $F_{\hat{A}\hat{B}}(t,t)=F_{\hat{A}\hat{B}}(0)=\langle\hat{A}(0)\hat{B}(0)\rangle$. Indeed, as the spur is invariant under cyclic permutations, we have

$$Sp\{\hat{A}(t)\hat{B}(t')e^{-\hat{\mathcal{H}}/\kappa_B T}\} = Sp\{e^{i\hat{\mathcal{H}}t}\hat{A}(0)e^{-i\hat{\mathcal{H}}(t-t')}\hat{B}(0)e^{-i\hat{\mathcal{H}}t'}e^{-\hat{\mathcal{H}}/\kappa_B T}\} = \tag{8-2}$$

$$= Sp\{\hat{A}(0)e^{-\hat{\mathcal{H}}(t-t')}\hat{B}(0)e^{\hat{\mathcal{H}}(t-t')}e^{-\hat{\mathcal{H}}/\kappa_B T}\},$$

whence it appears that $F_{\hat{A}\hat{B}}(t,t')=F_{\hat{A}\hat{B}}(t-t')$.

If eigenfunctions C_μ and eigenvalues $\mathcal{E}_\mu$ of the operator of Hamilton, $\hat{\mathcal{H}}$, are known, formulas (8-1) can be rewritten in the following form

$$\langle\hat{A}(t)\hat{B}(t')\rangle = Q^{-1}\sum_\mu (C^*_\mu \hat{A}(t)\hat{B}(t')C_\mu)e^{-\mathcal{E}_\mu/\kappa_B T} =$$

$$= Q^{-1}\sum_{\mu,\nu}(C^*_\mu \hat{A}(t)C_\nu)(C^*_\nu \hat{B}(t')C_\mu)e^{-\mathcal{E}_\mu/\kappa_B T}, \tag{8-3}$$

where $Q=\sum_\mu \exp(-\mathcal{E}_\mu/k_B T)$. Inserting into (8-3) the explicit form of time dependence of operators $\hat{A}(t)$ and $\hat{B}(t')$ according to formula (7-6) and taking into account that $\exp(i\hat{\mathcal{H}}t)C_\mu=\exp(i\mathcal{E}_\mu t)C_\mu$, we obtain

$$\langle\hat{A}(t)\hat{B}(t')\rangle = Q^{-1}\sum_{\mu,\nu}(C^*_\mu \hat{A}(0)C_\nu)(C^*_\nu \hat{B}(0)C_\mu)e^{-\mathcal{E}_\mu/\kappa_B T}e^{-i(\mathcal{E}_\mu-\mathcal{E}_\nu)(t'-t)} =$$

$$= Q^{-1}\sum_{\mu,\nu}(C^*_\mu \hat{A}(0)C_\nu)(C^*_\nu \hat{B}(0)C_\mu)e^{-\mathcal{E}_\nu/\kappa_B T}e^{(\mathcal{E}_\nu-\mathcal{E}_\mu)/\kappa_B T}e^{-i(\mathcal{E}_\nu-\mathcal{E}_\mu)(t-t')}. \tag{8-4}$$

Let us denote

$$J_{\hat{A}\hat{B}}(\omega) = Q^{-1}\sum_{\mu,\nu}(C^*_\mu \hat{A}(0)C_\nu)(C^*_\nu \hat{B}(0)C_\mu)e^{-\mathcal{E}_\mu/\kappa_B T}\delta(\mathcal{E}_\nu-\mathcal{E}_\mu-\omega). \tag{8-5}$$

Then eq. (8-4) can be rewritten in the following simple form

$$\langle \hat{A}(t)\hat{B}(t')\rangle = \int_{-\infty}^{+\infty} \mathcal{J}_{\hat{A}\hat{B}}(\omega)\, e^{\omega/\kappa_B T}\, e^{-i\omega(t-t')}\, d\omega . \tag{8-6}$$

The function $\mathcal{J}_{\hat{A}\hat{B}}(\omega)$ determined by eq. (8-5) is named the spectral representation or spectral intensity of the time correlation function $\langle\hat{A}(t)\hat{B}(t')\rangle$. The reader will see in the paragraphs to follow that in many cases it is more convenient to use just this representation.

Using the same approach we can find the spectral representation for the function $\langle\hat{B}(t')\hat{A}(t)\rangle$

$$\langle \hat{B}(t')\hat{A}(t)\rangle = \int_{-\infty}^{+\infty} \mathcal{J}_{\hat{A}\hat{B}}(\omega)\, e^{-i\omega(t-t')}\, d\omega . \tag{8-7}$$

To obtain the equation of evolution for the correlation function (8-1), let us differentiate eq. (8-1) over one of the two arguments of the function (for example, over the first one). Given in the right side of the new equation we obtain the products of a larger number of operators than in the initial eq. (8-1), an infinite chain of coupled equations of higher degrees is appeared

$$i\frac{d}{dt}\langle \hat{A}(t)\hat{B}(t')\rangle = \langle [\hat{A}(t),\hat{\mathcal{H}}]_{-}\,\hat{B}(t')\rangle,$$

$$i\frac{d}{dt}\langle [\hat{A}(t),\hat{\mathcal{H}}]_{-}\,\hat{B}(t')\rangle = \langle [[\hat{A}(t),\hat{\mathcal{H}}]_{-},\hat{\mathcal{H}}]_{-}\,\hat{B}(t')\rangle, \tag{8-8}$$

. . . .

The structure of the chain of eqs. (8-8) is very close to that of the chain of equations for classical correlation functions obtained by Bogolubov [25].

In order to calculate the macroscopic characteristics of the system we must know how to calculate the correlation functions. For example, to obtain the average of the operator $\hat{\mathcal{H}}$ which determines the energy of the system, we must be able to calculate the correlation functions $\langle \hat{P}^{+}_{n} \hat{P}_{n} \rangle$, $\langle \hat{P}^{+}_{n} \hat{P}_{m} \rangle$, $\langle \hat{P}^{+}_{n} \hat{P}^{+}_{m} \rangle$ and so on. Correlation functions are widely used in the problems of scattering, in studying the effects of point impurity and others. In the next Part of the book the reader will meet many applications of correlation functions.

§9. Method of Green functions.

In this paragraph we shall consider one of the most effective methods of calculation of the correlation functions – the method of Green functions.

In quantum statistical physics, two-times delayed ($G^{(a)}$) and passed ahead ($G^{(r)}$) Green functions are determined by the equations

$$G^{(r)}_{\hat{A}\hat{B}}(t,t') = \langle\langle \hat{A}(t)|\hat{B}(t')\rangle\rangle^{(r)} = \theta(t-t')\langle[\hat{A}(t),\hat{B}(t')]_{\eta}\rangle, \quad (9\text{-}1)$$

$$G^{(a)}_{\hat{A}\hat{B}}(t,t') = \langle\langle \hat{A}(t)|\hat{B}(t')\rangle\rangle^{(a)} = -\theta(t-t')\langle[\hat{A}(t),\hat{B}(t')]_{\eta}\rangle, \quad (9\text{-}2)$$

where $[\hat{A}(t),\hat{B}(t')] = \hat{A}(t)\hat{B}(t') - \eta\hat{B}(t')\hat{A}(t)$; η takes the values 1 or −1 (the choice of the value is arbitrary and does not depend on the commutation rules for operators $\hat{A}$ and $\hat{B}$). Usually, the value $\eta=1$ is taken when $\hat{A}$ and $\hat{B}$ are Bose-operators or any combinations of them, and the value $\eta=-1$ is taken when $\hat{A}$ and $\hat{B}$ are Fermi-operators or their combinations. $\theta(x)$ is the function of Heavyside. It is defined in the following way

$$\theta(x) = \begin{cases} 1, & \text{if } x > 0, \\ 0, & \text{if } x < 0. \end{cases} \quad (9\text{-}3)$$

As the Green functions (9-1) – (9-2) are determined by correlation functions, so we can expect that they will also depend only on the difference of the time arguments (as in the case of the correlation functions considered in the previous paragraph)

$$G^{(j)}_{\hat{A}\hat{B}}(t,t') = G^{(j)}_{\hat{A}\hat{B}}(t-t'), \quad j=r,a. \tag{9-4}$$

However, in contrast to the case of correlation functions, the Green functions are not determined at the point of coincidence of the time arguments t=t', which can be explained by the existence of the discontinuous factor $\Theta(t-t')$.

Let us construct now a chain of dynamical equations for Green functions. For this purpose let us differentiate with respect to time eqs. (9-1) and (9-2)

$$\frac{d}{dt}\langle\langle \hat{A}(t)|\hat{B}(t')\rangle\rangle^{(j)} = \frac{d\Theta(t-t')}{dt}\langle[\hat{A}(t),\hat{B}(t')]_\eta\rangle + \tag{9-5}$$
$$+\langle\langle \frac{d\hat{A}(t)}{dt}|\hat{B}(t')\rangle\rangle^{(j)}, \quad j=r,a.$$

Taking into account that

$$\frac{d}{dt}\Theta(t) = \delta(t), \tag{9-6}$$

and using the chain of coupled equations (7-9) for the operators, we obtain an infinite chain of coupled equations for the Green functions

$$i\frac{d}{dt}\langle\langle \hat{A}(t)|\hat{B}(t')\rangle\rangle^{(j)} = i\delta(t-t')\langle[\hat{A}(t),\hat{B}(t')]_\eta\rangle +$$
$$+\langle\langle[\hat{A}(t),\hat{\mathcal{H}}]_-|\hat{B}(t')\rangle\rangle^{(j)},$$

$$i\frac{d}{dt}\langle\langle[\hat{A}(t),\hat{\mathcal{H}}]_-|\hat{B}(t')\rangle\rangle^{(j)} = i\delta(t-t')\langle[[\hat{A}(t),\hat{\mathcal{H}}]_-,\hat{B}(t')]_\eta\rangle + \tag{9-7}$$
$$+\langle\langle[[\hat{A}(t),\hat{\mathcal{H}}]_-,\hat{\mathcal{H}}]_-|\hat{B}(t')\rangle\rangle^{(j)},$$

If we determine the Fourier-transforms (over time) of the Green functions by the equations

$$G^{(j)}_{\hat{A}\hat{B}}(t-t') = \langle\langle \hat{A}(t) | \hat{B}(t') \rangle\rangle^{(j)} = \int_{-\infty}^{+\infty} G^{(j)}_{\hat{A}\hat{B}}(E) e^{-iE(t-t')} dE, \qquad (9\text{-}8)$$

$$G^{(j)}_{\hat{A}\hat{B}}(E) = \langle\langle \hat{A} | \hat{B} \rangle\rangle^{(j)}_E = \frac{1}{2\pi} \int_{-\infty}^{+\infty} G^{(j)}_{\hat{A}\hat{B}}(t) e^{iEt} dt, \quad j = r, a, \qquad (9\text{-}9)$$

it is not difficult to obtain then a chain of the equations for Fourier-transforms of the Green functions, which is analogous to (9-7)

$$E \langle\langle \hat{A} | \hat{B} \rangle\rangle^{(j)}_E = \frac{i}{2\pi} \langle [\hat{A}, \hat{B}]_\eta \rangle + \langle\langle [\hat{A}, \hat{\mathcal{H}}] | \hat{B} \rangle\rangle^{(j)}_E,$$

$$E \langle\langle [\hat{A}, \hat{\mathcal{H}}] | \hat{B} \rangle\rangle^{(j)}_E = \frac{i}{2\pi} \langle [[\hat{A}, \hat{\mathcal{H}}], \hat{B}]_\eta \rangle + \langle\langle [[\hat{A}, \hat{\mathcal{H}}], \hat{\mathcal{H}}] | \hat{B} \rangle\rangle^{(j)}_E, \qquad (9\text{-}10)$$

. . . .

where it is taken into account that $\delta(t) = \frac{1}{2\pi} \int_{-\infty}^{+\infty} e^{-i\omega t} d\omega$.

To decouple the chain of eqs. (9-1), different approximate methods are used. However, in some particular cases when the model hamiltonian $\hat{\mathcal{H}}$ has some special form, the chain of the equations for the Fourier-transforms of the Green functions takes a closed form, and hence we can solve it exactly. The reader will meet examples of this type in the next Part.

Let us suppose that we succeeded in finding the Fourier-transforms of the Green function $G^{(r)}_{\hat{A}\hat{B}}(E)$. How can we relate it now with the corresponding correlation function $F_{\hat{A}\hat{B}}(t-t')$ or with its spectral density $\mathcal{J}_{\hat{A}\hat{B}}(\omega)$?

I hope that the reader remembers, that according to

definitions (9-1) and (9-2)

$$G^{(r)}_{\hat{A}\hat{B}}(E) = \frac{1}{2\pi}\int_{-\infty}^{+\infty} dt\, e^{iE(t-t')}\,\Theta(t-t')\langle \hat{A}(t)\hat{B}(t') - \eta\,\hat{B}(t')\hat{A}(t)\rangle. \qquad (9\text{-}11)$$

Inserting (8-6) and (8-7) into (9-11), we obtain

$$G^{(r)}_{\hat{A}\hat{B}}(E) = \frac{1}{2\pi}\int_{-\infty}^{+\infty} d\omega\, J_{\hat{A}\hat{B}}(\omega)(e^{\omega/\kappa_B T} - \eta)\int_{-\infty}^{+\infty} dt\, e^{i(E-\omega)t}\,\Theta(t). \qquad (9\text{-}12)$$

Let us take into account that the discontinuous function $\Theta(t)$ can be rewritten in the form of a contour integral

$$\Theta(t) = \frac{i}{2\pi}\int_{-\infty}^{+\infty} dE\, \frac{e^{-iEt}}{E + i\varepsilon} \qquad (\varepsilon > 0). \qquad (9\text{-}13)$$

The contours of integration used in formula (9-13) are shown in Fig. 13. With the help of (9-13) we find the integral over t in eq. (9-12)

$$\int_{-\infty}^{+\infty} dt\, e^{i(E-\omega)t}\,\Theta(t) = \frac{i}{E - \omega + i\varepsilon}. \qquad (9\text{-}14)$$

As a result, we obtain the following expression for the Fourier-transform of the delayed Green function

$$G^{(r)}_{\hat{A}\hat{B}}(E) = \frac{i}{2\pi}\int_{-\infty}^{+\infty} (e^{\omega/\kappa_B T} - \eta)\, J_{\hat{A}\hat{B}}(\omega)\, \frac{d\omega}{E - \omega + i\varepsilon} \qquad (9\text{-}15)$$

Using the same method we can obtain the Fourier-transform of passed ahead Green function

$$G^{(a)}_{\hat{A}\hat{B}}(E) = \frac{i}{2\pi}\int_{-\infty}^{+\infty} (e^{\omega/\kappa_B T} - \eta)\, \mathcal{J}_{\hat{A}\hat{B}}(\omega)\, \frac{d\omega}{E-\omega-i\varepsilon} . \tag{9-16}$$

If we consider E as a complex variable, formulas (9-15) and (9-16) can be combined in the following way

$$G_{\hat{A}\hat{B}}(E) = \frac{i}{2\pi}\int_{-\infty}^{+\infty} (e^{\omega/\kappa_B T} - \eta)\, \mathcal{J}_{\hat{A}\hat{B}}(\omega)\, \frac{d\omega}{E-\omega} =$$

$$= \begin{cases} G^{(r)}_{\hat{A}\hat{B}}(E), & \text{if } Im\, E > 0 , \\ G^{(a)}_{\hat{A}\hat{B}}(E), & \text{if } Im\, E < 0 . \end{cases} \tag{9-17}$$

The functions $G^{(r)}_{\hat{A}\hat{B}}(E)$ and $G^{(a)}_{\hat{A}\hat{B}}(E)$ are analytic in the upper and lower semi-surfaces, and can be considered as a unique analytical function $G_{\hat{A}\hat{B}}(E)$ having singularities (poles, lines of section) on the real axis. Formulas (9-15) - (9-17) are named the spectral representation (or spectral density) of the Green functions.

If the function $G_{\hat{A}\hat{B}}(E)$ is known, the spectral intensity $\mathcal{J}_{\hat{A}\hat{B}}(\omega)$ can be recovered by the following formula

$$(e^{\omega/\kappa_B T} - \eta)\, \mathcal{J}_{\hat{A}\hat{B}}(\omega) = G_{\hat{A}\hat{B}}(\omega + i\varepsilon) - G_{\hat{A}\hat{B}}(\omega - i\varepsilon) . \tag{9-18}$$

Indeed, let us take into account the known formula for delta-function

$$\frac{1}{x \pm i\varepsilon} = P\frac{1}{x} \mp i\pi\,\delta(x), \tag{9-19}$$

(here P symbolizes the principal value) and formula (9-17). Then it is easy to make sure that equation (9-18) is satisfied identically

$$G_{\hat{A}\hat{B}}(\omega+i\varepsilon) - G_{\hat{A}\hat{B}}(\omega-i\varepsilon) =$$

$$= \frac{i}{2\pi}\int_{-\infty}^{+\infty}(e^{\omega'/\kappa_B T}-\eta)\, J_{\hat{A}\hat{B}}(\omega')\left(\frac{1}{\omega-\omega'+i\varepsilon}-\frac{1}{\omega-\omega'-i\varepsilon}\right)d\omega' = \qquad (9\text{-}20)$$

$$= \int_{-\infty}^{+\infty}(e^{\omega'/\kappa_B T}-\eta)\, J_{\hat{A}\hat{B}}(\omega')\,\delta(\omega-\omega')\,d\omega' = (e^{\omega/\kappa_B T}-\eta)\, J_{\hat{A}\hat{B}}(\omega).$$

Once the spectral intensity $J_{\hat{A}\hat{B}}(\omega)$ has been found, calculations of correlation functions according to formulas (8-6) and (8-7) become possible.

PART 3. APPLICATIONS. EXCITED STATES OF BIOPOLYMERS.

In the previous Part the main ideas and methods of quantum statistical physics have been introduced. In this Part applications of the theory to the study of the low-lying excited states of biopolymers are considered. The problems of calculation of elementary excitations (excitons) spectra, correlation functions and macroscopic parameters characterizing excited states of biopolymers (their energy, density of excited states and others), are among them. We consider also applications of the problems to neutron scattering and the effects of irregularities of biopolymer structure. The main calculations are carried out for alpha-helical proteins. To make the calculations more clear and simple we show preference to the simplest models of alpha-helical proteins - molecular chains. However we give, where it is necessary, additional comments on the methods by which some more complex models could be investigated. Most of the results described in this Part have been obtained by the author and her colleagues in a series of works [26-31].

§10. Excited states of a chain consisting of identical molecules.

In this paragraph we shall investigate excited states of a one-dimensional molecular chain, which is the simplest model of biopolymer (Fig. 14). In this model the role of "molecules" is played by monomers (i.e. by peptide groups in the case of proteins, and by nucleotides in the case of nucleic acids). Our aim is to calculate the spectrum of elementary excitations and the correlation functions determining the main macroscopic parameters of the system.

Construction of the model hamiltonian. Let us assume that the arrangement of the energy levels of one isolated "molecule" is known (see, for example, Fig. 12). To simplify the calculations let us make two more assumptions:

1) suggest that the radius of the action of interaction forces does not exceed the distance between neighboring "molecules". This approximation is named the approximation of the nearest neighbors;

2) confine ourselves by consideration of only two lowest energy levels: the ground (0) and the first excited (1) (see terms E_0 and E_1 in Fig. 12).

The reader who attentively read Part 2 can easily conclude that the model system described above can be modeled by hamiltonian of type (5-23) with only one distinction: index m takes the value n+1 and n-1

$$\hat{\mathcal{H}} = \sum_n \Delta \hat{P}_n^+ \hat{P}_n + \sum_n \mathcal{M} (\hat{P}_n^+ \hat{P}_{n+1} + \hat{P}_{n+1}^+ \hat{P}_n) + \sum_n \mathcal{M}' (\hat{P}_n^+ \hat{P}_{n+1}^+ + \hat{P}_n \hat{P}_{n+1}) + \sum_n \mathcal{L} \hat{P}_n^+ \hat{P}_n \hat{P}_{n+1}^+ \hat{P}_{n+1} . \tag{10-1}$$

If we take into account that the number of elementary excitation is small, the last term in (10-1) describing interactions between elementary excitations, can be neglected. As a result we obtain the unknown hamiltonian in the form

$$\hat{\mathcal{H}} = \sum_n \Delta \hat{P}_n^+ \hat{P}_n + \sum_n \mathcal{M} (\hat{P}_n^+ \hat{P}_{n+1} + \hat{P}_{n+1}^+ \hat{P}_n) + \sum_n \mathcal{M}' (\hat{P}_n^+ \hat{P}_{n+1}^+ + \hat{P}_n \hat{P}_{n+1}) . \tag{10-2}$$

Boundary conditions are assumed to have the standard form $\hat{P}_n = \hat{P}_{n+N}$, $\hat{P}^+_n = \hat{P}^+_{n+N}$.

Calculation of the spectrum of elementary excitations by the method of diagonalization of the model hamiltonian. The technique of calculations of spectra in terms of Pauli-operators is not well-developed, and it is more convenient to rewrite hamiltonian (10-2) in terms of Bose- (or Fermi-)operators and then use the well known technique of calculation for Bose- (or Fermi-) systems to work out the results.

However, an attempt to rewrite hamiltonian (10-2) in terms of Bose-operators by using the exact formula (6-10) immediately leads us to the substantial complication of the hamiltonian

$$\hat{\mathcal{H}} = \sum_n \sum_{\nu,\mu} \frac{(-2)^{(\nu+\mu)/2}}{\sqrt{(1+\nu)!\,(1+\mu)!}} \Big\{ \Delta \hat{b}_n^+ (\hat{b}_n^{+\nu} \hat{b}_n^{\nu})^{1/2} (\hat{b}_n^{+\mu} \hat{b}_n^{\mu})^{1/2} \hat{b}_n + \mathcal{M} \big[\hat{b}_n^+ (\hat{b}_n^{+\nu} \hat{b}_n^{\nu})^{1/2} (\hat{b}_{n+1}^{+\mu} \hat{b}_{n+1}^{\mu})^{1/2} \hat{b}_{n+1} + \hat{b}_{n+1}^+ (\hat{b}_{n+1}^{+\nu} \hat{b}_{n+1}^{\nu})^{1/2} (\hat{b}_n^{+\mu} \hat{b}_n^{\mu})^{1/2} \hat{b}_n \big] + \mathcal{M}' \big[\hat{b}_n^+ (\hat{b}_n^{+\nu} \hat{b}_n^{\nu})^{1/2} (\hat{b}_{n+1}^+) (\hat{b}_{n+1}^{+\mu} \hat{b}_{n+1}^{\mu})^{1/2} + (\hat{b}_n^{+\nu} \hat{b}_n^{\nu})^{1/2} (\hat{b}_n) (\hat{b}_{n+1}^{+\mu} \hat{b}_{n+1}^{\mu})^{1/2} \hat{b}_{n+1} \big] \Big\}. \qquad (10\text{-}3)$$

So, it remains only to try to rewrite (10-2) in terms of Fermi-operators. For this purpose let us use eqs. (6-16). After inserting (6-16) into (10-2) and taking into account commutation rules (6-18) - (6-19) we obtain

$$\hat{\mathcal{H}} = \sum_n \Delta \hat{a}_n^{+} \hat{a}_n + \sum_n \mathcal{M} (\hat{a}_n^{+} \hat{a}_{n+1} + \hat{a}_{n+1}^{+} \hat{a}_n) + \sum_n \mathcal{M}' (\hat{a}_n^{+} \hat{a}_{n+1}^{+} - \hat{a}_n \hat{a}_{n+1}). \tag{10-4}$$

Thus, transformation of Pauli-operators to Fermi-ones does not lead to any complication of the structure of the initial hamiltonian. This is a remarkable result because it gives us an opportunity to solve the problem exactly.

The solution can be obtained in several ways. For example, we can take into account that hamiltonian (10-4) is a positive definite form: then, according to theorems of linear algebra, it can be diagonalized by some unitary transformation. Indeed, let us choose the transformation in the most general form

$$\hat{a}_n = \frac{1}{\sqrt{N}} \sum_k \{ u_k \hat{\alpha}_k - v_{-k}^{*} \hat{\alpha}_{-k}^{+} \} e^{ikn},$$
$$\hat{a}_n^{+} = \frac{1}{\sqrt{N}} \sum_k \{ u_k^{*} \hat{\alpha}_k + v_{-k} \hat{\alpha}_{-k} \} e^{-ikn}, \tag{10-5}$$

where k is a wave vector, laying in the first Brillouin zone; $\hat{\alpha}_k^{+}$, $\hat{\alpha}_k$ are Fermi-operators; u_k, v_k are the coefficients of the linear transformation, which satisfy the following conditions

$$u_k = u_k^{*}, \quad u_k = u_{-k}, \quad v_k = -v_k^{*}, \quad v_k = -v_{-k},$$
$$u_k u_{-k}^{*} - v_k v_{-k}^{*} = 1. \tag{10-6}$$

Last equality in (10-6) is associated with the requirements of the unitarity.

Let us insert (10-5) into hamiltonian (10-4) and require the cross terms to be equal to zero

$$(\Delta + 2\mathcal{M} \cos ka)v_{-k}u_k + i\mathcal{M}' \sin ka (v_{-k}v_k - u_{-k}u_k) = 0$$

$$(\Delta + 2\mathcal{M} \cos ka)u_k^* v_{-k}^* + i\mathcal{M}' \sin ka (u_k^* u_{-k}^* - v_k^* v_{-k}^*) = 0. \quad (10\text{-}7)$$

Then the model hamiltonian appears the diagonalized form

$$\hat{\mathcal{H}} = Const. + \sum_k E(k)\hat{\alpha}_k^+ \hat{\alpha}_k , \quad (10\text{-}8)$$

where E(k) is just the unknown spectrum of elementary excitations

$$E(k) = \{(\Delta + 2\mathcal{M} \cos ka)^2 + (2\mathcal{M}' \sin ka)^2\}^{1/2}, \quad (10\text{-}9)$$

and the coefficients u_k, v_k are determined by eqs. (10-6) and (10-7)

$$u_k^2 = \frac{1}{2}\{1 + (\Delta + 2\mathcal{M} \cos ka)/E(k)\},$$

$$|v_k|^2 = \frac{1}{2}\{1 - (\Delta + 2\mathcal{M} \cos ka)/E(k)\}. \quad (10\text{-}10)$$

What follows from these results? First of all eq. (10-8) indicates that elementary excitations (excitons) of the molecular chain considered obey Fermi-statistics, and in the frame of approximation of low density of excitations they meet the requirements of the model of the ideal gas of fermions with spectrum (10-9). This spectrum considerably differs from that of one isolated "molecule". Thus, in this low-density of excitations approximation, instead of the term (1) in the spectrum of one isolated "molecule", a band (or a zone) takes place in the spectrum of the chain. The width of the band is determined by the difference: $E_{max}(k)-E_{min}(k)$, and the density of the excited states in the band by the sum: $\sum_K \langle \hat{\alpha}^+_k \hat{\alpha}_k \rangle$. Dispersion equation (10-9) permits us to find (i) the group velocity of the movement of elementary excitations along the molecular chain: $v=\partial E(k)/\partial k$, and (ii) the time of the transmission of excitation from one molecule to the next (neighboring) molecule: $\tau=a/v$. Using the model values $|M|\cong|M'|\leqslant 10^3\ cm^{-1}$, $a\cong 5\cdot 10^{-8}$ cm, which are typical for molecular crystal lattices, we can find upper estimation of the group velocity: $|v|<2a|M|\cong 5\cdot 10^5$ m/sec, and lower one of the time of transmission: $\tau=a/v>10^{-3}$ picosec.

Calculation of the spectrum of elementary excitations by the method of Green functions. The problem (10-4) can be solved by another method which is known as the method of Green functions. To determine the types of Green functions required and to construct the dynamical equations for them, let us write the equations for operators $\hat{a}^+_n$, $\hat{a}_n$ in Heisenberg representation. These equations can be found by substituting $\hat{a}_n$ and $\hat{a}^+_n$ for $\hat{A}$ in eq. (7-9) with hamiltonian (10-4). After substituting we obtain

$$i\frac{d\hat{a}_n}{dt} = \Delta\hat{a}_n + M(\hat{a}_{n+1}+\hat{a}_{n-1}) + M'(\hat{a}^+_{n+1}-\hat{a}^+_{n-1}),$$
$$i\frac{d\hat{a}^+_n}{dt} = -\Delta\hat{a}^+_n - M(\hat{a}^+_{n+1}+\hat{a}^+_{n-1}) - M'(\hat{a}_{n+1}-\hat{a}_{n-1}). \qquad (10\text{-}11)$$

The form of eqs. (10-11) suggests to introduce the following Green functions

$$G^{(1)}_{n-m}(t-t') = \langle\langle \hat{a}_n(t) | \hat{a}^+_m(t') \rangle\rangle,$$

$$G^{(2)}_{n-m}(t-t') = \langle\langle \hat{a}^+_n(t) | \hat{a}^+_m(t') \rangle\rangle. \qquad (10\text{-}12)$$

The dependence of functions (10-12) on the difference of indexes (n-m) is explained by the property of translational invariance of the model system. Taking into account this property, we introduce the Fourier-expansion of function (10-12) not only in time (according to formula (9-8)) but also in the variable z=na

$$G^{(j)}_{n-m}(t-t') = \frac{1}{N}\sum_K \int G^{(j)}_K(E)\, e^{iKa(n-m)}\, e^{-iE(t-t')}\, dE, \quad j=1,2. \qquad (10\text{-}13)$$

Let us construct equations of motions for the Green functions (10-12) according to formulas (9-7) and expand them according to rule (10-13). As a result we obtain

$$(E-\Delta)\, G^{(1)}_K(E) = \frac{i}{2\pi} + 2M \cos Ka\; G^{(1)}_K(E) + 2iM' \sin Ka\; G^{(2)}_K(E),$$

$$(E+\Delta)\, G^{(2)}_K(E) = -2M \cos Ka\; G^{(2)}_K(E) - 2iM' \sin Ka\; G^{(1)}_K(E). \qquad (10\text{-}14)$$

Here we used the known formula of solid state theory

$$\delta(n-m) = \frac{1}{N}\sum_K e^{iKa(n-m)}. \qquad (10\text{-}15)$$

From the condition of existence of non-trivial solution of the system of linear algebraic eqs. (10-14)

$$\begin{vmatrix} (E-\Delta-2\mathcal{M}\cos ka) & -2i\mathcal{M}'\sin ka \\ 2i\mathcal{M}'\sin ka & (E+\Delta+2\mathcal{M}\cos ka) \end{vmatrix} = 0 \qquad (10\text{-}16)$$

we obtain the spectrum of elementary excitations

$$E(k) = \{(\Delta + 2\mathcal{M}\cos ka)^2 + (2\mathcal{M}'\sin ka)^2\}^{1/2},$$

As it could be expected, this result coincides with formula (10-9). Solving at last the system of eqs. (10-14) we find the Fourier-components of Green functions

$$G_k^{(1)}(E) = \frac{i}{2\pi}\,\frac{E+\Delta+2\mathcal{M}\cos ka}{E^2-E^2(k)},$$

$$G_k^{(2)}(E) = \frac{1}{\pi}\,\frac{\mathcal{M}'\sin ka}{E^2-E^2(k)}. \qquad (10\text{-}17)$$

To calculate different correlation functions it is very useful to know another two Green functions: $G^{(3)}_{n-m}(t-t') = \langle\langle \hat{a}_n(t) | \hat{a}_m(t')\rangle\rangle$ and $G^{(4)}_{n-m}(t-t') = \langle\langle \hat{a}^+_n(t) | \hat{a}_m(t')\rangle\rangle$. Following the scheme of calculations carried out above we can easily obtain corresponding results for the Fourier-components $G^{(3)}_k$ and $G^{(4)}_k$

$$G_k^{(3)}(E) = -\frac{1}{\pi}\,\frac{\mathcal{M}'\sin ka}{E^2-E^2(k)},$$

$$G_k^{(4)}(E) = \frac{i}{2\pi}\,\frac{E-\Delta-2\mathcal{M}\cos ka}{E^2-E^2(k)}. \qquad (10\text{-}18)$$

Calculation of correlation functions. It remains only to show now the method of calculations of different correlation functions. For example, let us consider a scheme of calculation of the correlation function $\langle \hat{a}^{+}_{n} \hat{a}_{n} \rangle$ which determines the density of excitations in the n-th point of one-dimensional lattice (or chain). For this purpose let us rewrite the Green function $G^{(1)}_{k}(E)$ (see formula (10-17)) in the following equivalent form

$$G^{(1)}_{\kappa}(E)=\frac{i}{4\pi}\left\{\left(1+\frac{T_{\kappa}}{E(\kappa)}\right)\frac{1}{E-E(\kappa)}+\left(1-\frac{T_{\kappa}}{E(\kappa)}\right)\frac{1}{E+E(\kappa)}\right\}, \qquad (10\text{-}19)$$

where $T_{k}=\Delta+2M\cos ka$. Insert (10-19) into formula (9-18) and take into account eq. (9-19). Then we obtain the following equation for the spectral intensity $\mathcal{J}^{(1)}_{k}(\omega)$

$$\mathcal{J}^{(1)}_{\kappa}(\omega)=\frac{1}{2}\left(1+\frac{T_{\kappa}}{E(\kappa)}\right)\frac{\delta(\omega-E(\kappa))}{1+\exp(E(\kappa)/2\kappa_{B}T)}+$$

$$+\frac{1}{2}\left(1-\frac{T_{\kappa}}{E(\kappa)}\right)\frac{\delta(\omega+E(\kappa))}{1+\exp(-E(\kappa)/2\kappa_{B}T)}, \qquad (10\text{-}20)$$

whence it appears that the formula for the correlation function $\langle \hat{a}^{+}_{n} \hat{a}_{n} \rangle$ has the form

$$\langle \hat{a}^{+}_{n}\hat{a}_{n}\rangle=\langle \hat{P}^{+}_{n}\hat{P}_{n}\rangle=\frac{1}{N}\sum_{\kappa}\int_{-\infty}^{+\infty}\mathcal{J}^{(1)}_{\kappa}(\omega)\,d\omega=$$

$$=\frac{1}{2N}\sum_{\kappa}\left(1-\frac{T_{\kappa}}{E(\kappa)}\,\mathrm{th}\,\frac{E(\kappa)}{2\kappa_{B}T}\right). \qquad (10\text{-}21)$$

Note, that formula (10-21) has been obtained by inserting (10-20) into (8-7) when $t=t'$. In the general case ($t\neq t'$) we obtain

$$\langle \hat{a}^{+}_{n}(t)\,\hat{a}_{n}(t')\rangle=\frac{1}{N}\sum_{\kappa}\int_{-\infty}^{+\infty}\mathcal{J}^{(1)}_{\kappa}(\omega)\,e^{-i\omega(t-t')}\,d\omega=$$

$$= \frac{1}{2N}\sum_{k}\left\{\left(1+\frac{T_k}{E(k)}\right)\frac{\exp(-iE(k)(t-t'))}{1+\exp(E(k)/2\kappa_B T)} + \right. \quad (10\text{-}22)$$

$$\left. + \left(1-\frac{T_k}{E(k)}\right)\frac{\exp(iE(k)(t-t'))}{1+\exp(-E(k)/2\kappa_B T)}\right\}.$$

We can note a definite conformity with a law, which is typical for the problem considered: poles in eq. (10-17) for the Green function $G_k^{(1)}(E)$ lead to δ-form singularities in formula (10-20) for spectral intensity $\mathcal{J}_k^{(1)}(\omega)$ and to oscillations with the frequency $\omega=E(k)$* (we continue to suggest that $\hbar=1$) in formula (10-22) for correlation function.

Other correlation functions: $\langle\hat{a}^{+}_{n}\hat{a}_{n+1}\rangle$, $\langle\hat{a}^{+}_{n+1}\hat{a}_{n}\rangle$, $\langle\hat{a}^{+}_{n}\hat{a}^{+}_{n+1}\rangle$ and $\langle\hat{a}_{n}\hat{a}_{n+1}\rangle$ can be calculated by the analogous scheme. If we calculate these functions, we can obtain different macroscopic characteristics of the system. We can give here, for example, the result of the calculation for the energy of the system, $\mathcal{E}$,

$$\mathcal{E} = \langle\hat{\mathcal{H}}\rangle = \sum_{n}\{\Delta\langle\hat{P}^{+}_{n}\hat{P}_{n}\rangle + \mathcal{M}\langle\hat{P}^{+}_{n}\hat{P}_{n+1}+\hat{P}^{+}_{n+1}\hat{P}_{n}\rangle +$$

$$+\mathcal{M}'\langle\hat{P}^{+}_{n}\hat{P}^{+}_{n+1}+\hat{P}_{n}\hat{P}_{n+1}\rangle\} = \sum_{n}\{\Delta\langle\hat{a}^{+}_{n}\hat{a}_{n}\rangle +$$

$$+\mathcal{M}\langle\hat{a}^{+}_{n}\hat{a}_{n+1}+\hat{a}^{+}_{n+1}\hat{a}_{n}\rangle + \mathcal{M}'\langle\hat{a}^{+}_{n}\hat{a}^{+}_{n+1}-\hat{a}_{n}\hat{a}_{n+1}\rangle\} = \quad (10\text{-}23)$$

$$= \frac{1}{2}\sum_{k}\left\{T_k\left(1-\frac{T_k}{E(k)}\,\mathrm{th}\,\frac{E(k)}{2\kappa_B T}\right)+\frac{L_k}{E(k)}\,\mathrm{th}\,\frac{E(k)}{2\kappa_B T}\right\}.$$

where L_k =2M'sinka.

---

* Pay attention to the notations for the frequencies and the energies: they coincide here because we assume that $\hbar=1$.

§11. Excited states of a chain consisting of the molecules of two types.

In the previous paragraph the simplest physical model of biopolymers, which is a chain consisting of N identical "molecules" weakly interacting with one another, was considered. However, in the real systems the monomers forming the chains differ from one another and form quite definite sequences. In this paragraph we consider a simple example of such a sequence

A B A B A B A B. . .

From the physical point of view this model is equivalent to one-dimensional molecular crystal with two molecules in the cell (Fig. 15). To simplify the problem let us assume that the "molecules" of the chain differ from one another only by orientation and /or by distance between the nearest neighbors. In this case the arrangement of energy levels of one isolated "molecule" will be the same for all molecules of the chain.

Construction of the model hamiltonian. Using the hamiltonian of type (5-23) and following the logical scheme of the previous paragraph, it is easy to obtain the model hamiltonian for the chain consisting of two types "molecules"

$$\hat{\mathcal{H}} = \sum_n \Delta\,(\hat{P}^+_{n,1}\hat{P}_{n,1} + \hat{P}^+_{n,2}\hat{P}_{n,2}) + \sum_n m\,(\hat{P}^+_{n,1}\hat{P}_{n,2} + \hat{P}^+_{n,2}\hat{P}_{n,1}) +$$

$$+ \sum_n \mathcal{M}\,(\hat{P}^+_{n,2}\hat{P}_{n+1,1} + \hat{P}^+_{n+1,1}\hat{P}_{n,2}) + \sum_n m'(\hat{P}^+_{n,1}\hat{P}^+_{n,2} + \hat{P}_{n,1}\hat{P}_{n,2}) +$$

$$+ \sum_n \mathcal{M}'(\hat{P}^+_{n,2}\hat{P}^+_{n+1,1} + \hat{P}_{n,2}\,\hat{P}_{n+1,1}), \qquad (11\text{-}1)$$

where we introduced the following notation : $m = M_{n,1;n,2}$, $M = M_{n,2;n+1,1}$, $m' = M'_{n,1;n,2}$, $M' = M'_{n,2;n+1,1}$.

<u>Calculation of the spectrum of elementary excitations</u>. As we have done in §10, we shall begin with rewriting hamiltonian (11-1) in terms of Fermi-operators

$$\hat{\mathcal{H}} = \sum_n \Delta(\hat{a}^+_{n,1}\hat{a}_{n,1} + \hat{a}^+_{n,2}\hat{a}_{n,2}) + \sum_n m(\hat{a}^+_{n,1}\hat{a}_{n,2} + \hat{a}^+_{n,2}\hat{a}_{n,1}) +$$
$$+ \sum_n \mathcal{M}(\hat{a}^+_{n,2}\hat{a}_{n+1,1} + \hat{a}^+_{n+1,1}\hat{a}_{n,2}) + \sum_n m'(\hat{a}^+_{n,1}\hat{a}^+_{n,2} - \hat{a}_{n,1}\hat{a}_{n,2}) + \tag{11-2}$$
$$+ \sum_n \mathcal{M}'(\hat{a}^+_{n,2}\hat{a}^+_{n+1,1} - \hat{a}_{n,2}\hat{a}_{n+1,1}).$$

Let us construct then the equations of motions for the operators $\hat{a}_{n,1}$, $\hat{a}_{n,2}$, $\hat{a}^+_{n,1}$, $\hat{a}^+_{n,2}$

$$i\frac{d\hat{a}_{n,1}}{dt} = \Delta\hat{a}_{n,1} + m\hat{a}_{n,2} + \mathcal{M}\hat{a}_{n-1,2} + m'\hat{a}^+_{n,2} - \mathcal{M}'\hat{a}^+_{n-1,2},$$
$$i\frac{d\hat{a}^+_{n,1}}{dt} = -\Delta\hat{a}^+_{n,1} - m\hat{a}^+_{n,2} - \mathcal{M}\hat{a}^+_{n-1,2} - m'\hat{a}_{n,2} + \mathcal{M}'\hat{a}_{n-1,2}; \tag{11-3}$$
$$i\frac{d\hat{a}_{n,2}}{dt} = \Delta\hat{a}_{n,2} + m\hat{a}_{n,1} + \mathcal{M}\hat{a}_{n+1,1} - m'\hat{a}^+_{n,1} + \mathcal{M}'\hat{a}^+_{n+1,1},$$
$$i\frac{d\hat{a}^+_{n,2}}{dt} = -\Delta\hat{a}^+_{n,2} - m\hat{a}^+_{n,1} - \mathcal{M}\hat{a}^+_{n+1,1} + m'\hat{a}_{n,1} - \mathcal{M}'\hat{a}_{n+1,1}.$$

According to the structure of eqs.(11-3), introduce four types of Green functions

$$G^{(1)}_{n-m}(t-t') = \langle\langle \hat{a}_{n,1}(t) | \hat{a}^+_{m,1}(t') \rangle\rangle,$$
$$G^{(2)}_{n-m}(t-t') = \langle\langle \hat{a}^+_{n,1}(t) | \hat{a}^+_{m,1}(t') \rangle\rangle, \tag{11-4}$$
$$G^{(3)}_{n-m}(t-t') = \langle\langle \hat{a}_{n,2}(t) | \hat{a}^+_{m,1}(t') \rangle\rangle,$$
$$G^{(4)}_{n-m}(t-t') = \langle\langle \hat{a}^+_{n,2}(t) | \hat{a}^+_{m,1}(t') \rangle\rangle$$

and write equations of motion for Fourier-components of these functions, which are determined by formula (10-13)

$$(E-\Delta)G_K^{(1)} = \frac{i}{2\pi} + (\mathcal{M}e^{-ika}+m)G_K^{(3)} + (-\mathcal{M}'e^{-ika}+m')G_K^{(4)},$$
$$(E+\Delta)G_K^{(2)} = -(\mathcal{M}e^{-ika}+m)G_K^{(4)} + (\mathcal{M}'e^{-ika}-m')G_K^{(3)},$$
$$(E-\Delta)G_K^{(3)} = (\mathcal{M}e^{ika}+m)G_K^{(1)} + (\mathcal{M}'e^{ika}-m')G_K^{(2)}, \qquad (11\text{-}5)$$
$$(E+\Delta)G_K^{(4)} = -(\mathcal{M}e^{ika}+m)G_K^{(2)} + (-\mathcal{M}'e^{ika}+m')G_K^{(1)}.$$

From the condition of existence of non-trivial solution of the system of eqs. (11-5) we find the spectrum of elementary excitations

$$E_{1,2}(k) = \{\Delta^2 + t_K + t'_K \pm [4\Delta^2 t_K + (t_K + t'_K)^2 - q_K p_K]^{1/2}\}^{1/2}, \qquad (11\text{-}6)$$

where $t_K = \mathcal{M}^2 + m^2 + 2\mathcal{M}m\cos ka$, $t'_K = \mathcal{M}'^2 + m'^2 - 2\mathcal{M}'m'\cos ka$,

$g_K = (\mathcal{M}+\mathcal{M}')^2 + (m-m')^2 + 2(\mathcal{M}+\mathcal{M}')(m-m')\cos ka$,

$p_K = (\mathcal{M}-\mathcal{M}')^2 + (m+m')^2 + 2(\mathcal{M}-\mathcal{M}')(m+m')\cos ka$.

In the particular case, when $M \cong M'$, $m \cong m'$, the spectrum $E_{1,2}(k)$ has the form

$$E_{1,2}(k) = \{\Delta^2 + 2(\mathcal{M}^2+m^2) \pm 2[(\mathcal{M}^2+m^2+2\mathcal{M}m\cos ka)\Delta^2 + (\mathcal{M}^2-m^2)^2]^{1/2}\}^{1/2}. \qquad (11\text{-}7)$$

In the "degenerate" case, when $M \cong m$, $M' \cong m'$, the dispersion formula (11-6) takes the simpler form

$$E_{1,2}(k) = \{(\Delta \pm 2\mathcal{M}\cos(ka/2))^2 + (2\mathcal{M}'\sin(ka/2))^2\}^{1/2}. \qquad (11\text{-}8)$$

As one should expect, solution (11-8) taken with plus and with the substitution of $a/2$ for a (compare Fig. 14 and Fig. 15), coincides with formula (10-9) which determines the spectrum of elementary excitations in a chain consisting of identical "molecules".

The results obtained above show that the energy term (1) of the spectrum of one isolated "molecule" splits up into two bands (or zones): $E_1(k)$ and $E_2(k)$. The formation of the bands is explained by taking into consideration interactions between "molecules" and by the existence of two types of "molecules" (A- and B-type) in the lattice cell. Using the same formulas as in §10, we can calculate the width of the bands, the density of the excited states in each of the bands and the group velocity for the transmission of elementary excitations from one "molecule" to another. Besides these three parameters, the spectrum obtained is characterized by an additional value - the splitting, depending on the wave vector k: $\nu(k)=E_1(k)-E_2(k)$.

The reader can compare the spectral curves obtained for chains with one and two "molecules" in the lattice cell (Fig. 16). Calculations were carried out for the particular case when $M = M'$, $m=m'$, $a=1$ and for three different model values of the parameters Δ, M and m.

<u>Calculation of correlation functions.</u> According to §9 we can calculate different correlation functions of the system if the Green functions (11-4) are known. For example let us describe the scheme of calculation of the average number of excitations in the point {n, 1} of lattice: $\langle \hat{a}^{+}_{n,1} \hat{a}_{n,1} \rangle$.

First of all let us find the Fourier-component of the Green $G_k^{(1)}$ by the help of formulas (11-5)

$$G_K^{(1)}(E) = \frac{i}{2\pi} \frac{E^2-\Delta^2}{(E^2-E_1^2(K))(E^2-E_2^2(K))} \left\{ E+\Delta - \frac{t_K}{E+\Delta} - \frac{t'_K}{E-\Delta} \right\} \qquad (11\text{-}9)$$

and rewrite the obtained eq. (11-9) in the equivalent form

$$G_K^{(1)}(E) = \frac{i}{2\pi} \left\{ \left(1+\frac{T_K^{(1)}}{E_1(K)}\right) \frac{\nu_K^{(1)}}{E-E_1(K)} + \left(1-\frac{T_K^{(1)}}{E_1(K)}\right) \frac{\nu_K^{(1)}}{E+E_1(K)} - \right.$$
$$\left. -\left(1+\frac{T_K^{(2)}}{E_2(K)}\right)\frac{\nu_K^{(2)}}{E-E_2(K)} - \left(1-\frac{T_K^{(2)}}{E_2(K)}\right)\frac{\nu_K^{(2)}}{E+E_2(K)} \right\}, \qquad (11\text{-}10)$$

where

$$T_K^{(j)} = \Delta(\Delta^2 - t_K + t'_K - E_j^2(K))/(\Delta^2 + t_K + t'_K - E_j^2(K)),$$
$$\nu_K^{(j)} = (\Delta^2 + t_K + t'_K - E_j^2(K))/(E_2^2(K) - E_1^2(K)), \quad j=1,2.$$

If we use eqs. (9-18) and (9-19) we can find the spectral intensity $\mathcal{J}_k^{(1)}(\omega)$

$$\mathcal{J}_K^{(1)}(\omega) = \frac{\nu_K^{(1)}}{2} \left\{ \left(1+\frac{T_K^{(1)}}{E_1(K)}\right) \frac{\delta(\omega-E_1(K))}{1+\exp(E_1(K)/k_B T)} + \left(1-\frac{T_K^{(1)}}{E_1(K)}\right) \frac{\delta(\omega+E_1(K))}{1+\exp(-E_1(K)/k_B T)} \right\} -$$
$$- \frac{\nu_K^{(2)}}{2} \left\{ \left(1+\frac{T_K^{(2)}}{E_2(K)}\right) \frac{\delta(\omega-E_2(K))}{1+\exp(E_2(K)/k_B T)} + \right. \qquad (11\text{-}11)$$
$$\left. + \left(1-\frac{T_K^{(2)}}{E_2(K)}\right) \frac{\delta(\omega+E_2(K))}{1+\exp(E_2(K)/k_B T)} \right\}.$$

Then with the help of formula (8-7) we obtain the unknown average

$$\langle \hat{a}^+_{n,1} \hat{a}_{n,1} \rangle = \frac{1}{2N} \sum_K \left\{ \nu_K^{(1)} \left[1 - \frac{T_K^{(1)}}{E_1(K)} th\left(\frac{E_1(K)}{2k_B T}\right)\right] - \right.$$
$$\left. - \nu_K^{(2)} \left[1 - \frac{T_K^{(2)}}{E_2(K)} th\left(\frac{E_2(K)}{2k_B T}\right)\right] \right\}. \qquad (11\text{-}12)$$

In conclusion, we note that the method of calculation of spectra and correlation functions described in §10, 11 can be applied to other problems, which could be united by the common title : investigation of excited states of molecular chains with arbitrary number of "molecules" in the lattice cell. In other words, besides the sequences of types

A A A ... and

A B A B A B...

the method described above can be used to investigate the sequences of types

A B C A B C A B C...,

A B C D A B C D A B C D...

and any other regular sequences.

§12. Exciton spectrum of alpha-helical fragment of protein molecule.

The chemical formula and the structure of alpha-helical proteins were discussed in details in §2. A schematic picture of the approximate "physical" model of the molecule, proposed by Davydov and Suprun [32], is shown in Fig. 17. The peptide groups are denoted by circles, and the bonds between the groups - by lines. The main simplifications of the model are

1) it is assumed that the difference between the side groups $R_1, R_2, \ldots R_N$ is not take into account;
2) it is assumed that every turn of the helix contains 3 peptide groups instead of 3,6 as it takes happens in the real proteins;
3) it is assumed that the spectrum of any isolated peptide group can be limited by two energy levels: the ground (0) and first excited (1); as a first excited level, the characteristic band (Amid-1) having the energy 0,21 eV, is chosen;
4) only interactions between the nearest neighboring peptide groups (that is between the groups which are neighbours both along the helix and along the chains formed by hydrogen bonds) are taken into account.

It is convenient to renumerate the peptide groups by two indexes n and α, the index n numerating the peptide groups along one of the three chains formed by hydrogen bonds and the index α indicating the chain where the peptide group is placed. Now it must become obvious for the reader that the model of Davydov and

Suprun is equivalent to the model of molecular chain with three "molecules" in the lattice cell (Fig. 18). This model advantageously differs from the models considered in §10, 11, because it takes into account a more accurate information about the structure and interactions.

The problem which is brought up consists in studying the changes of the spectrum when we pass from the model of isolated "molecule" to the model described above.

Let us write the hamiltonian for the model of Davydov and Suprun

$$\hat{\mathcal{H}} = \sum_n \sum_\alpha \Delta \hat{P}^+_{n,\alpha} \hat{P}_{n,\alpha} + \sum_n \sum_\alpha \mathcal{M}_{n,\alpha;n+1,\alpha} (\hat{P}^+_{n,\alpha} \hat{P}_{n+1,\alpha} + \hat{P}^+_{n+1,\alpha} \hat{P}_{n,\alpha}) +$$
$$+ \sum_n \sum_\alpha \mathcal{M}'_{n,\alpha;n+1,\alpha} (\hat{P}^+_{n,\alpha} \hat{P}^+_{n+1,\alpha} + \hat{P}_{n,\alpha} \hat{P}_{n+1,\alpha}) + \sum_n \{\mathcal{M}_{n,1;n,2} (\hat{P}^+_{n,1} \hat{P}_{n,2} + \hat{P}^+_{n,2} \hat{P}_{n,1}) +$$
$$+ \mathcal{M}_{n,2;n,3} (\hat{P}^+_{n,2} \hat{P}_{n,3} + \hat{P}^+_{n,3} \hat{P}_{n,2}) + \mathcal{M}_{n,3;n+1,1} (\hat{P}^+_{n,3} \hat{P}_{n+1,1} + \hat{P}^+_{n+1,1} \hat{P}_{n,3})\} +$$
$$+ \sum_n \{\mathcal{M}'_{n,1;n,2} (\hat{P}^+_{n,1} \hat{P}^+_{n,2} + \hat{P}_{n,1} \hat{P}_{n,2}) + \mathcal{M}'_{n,2;n,3} (\hat{P}^+_{n,2} \hat{P}^+_{n,3} + \hat{P}_{n,2} \hat{P}_{n,3}) + \qquad (12\text{-}1)$$
$$+ \mathcal{M}'_{n,3;n+1,1} (\hat{P}^+_{n,3} \hat{P}^+_{n+1,1} + \hat{P}_{n,3} \hat{P}_{n+1,1}),$$

where $M_{n,\alpha;n+1,\alpha}$, $M'_{n,\alpha;n+1,\alpha}$ are the matrix elements of the operator of interaction between the peptide groups which are the nearest along the hydrogen chains; $M_{n,\alpha;n,\beta}$, $M'_{n,\alpha;n,\beta}$ are the matrix elements of the operator of interaction between the peptide groups which are the nearest along the helix ($\alpha,\beta = 1, 2, 3;\ \alpha \neq \beta$).

Let us rewrite hamiltonian (12-1) in terms of the Fermi-operators

$$\hat{\mathcal{H}} = \sum_n \sum_\alpha \Delta \hat{a}^+_{n,\alpha} \hat{a}_{n,\alpha} + \sum_n \sum_\alpha \mathcal{M} (\hat{a}^+_{n,\alpha} \hat{a}_{n+1,\alpha} + \hat{a}^+_{n+1,\alpha} \hat{a}_{n,\alpha}) +$$

$$+\sum_n\sum_\alpha \mathcal{M}'(\hat{a}^+_{n,\alpha}\hat{a}^+_{n+1,\alpha}-\hat{a}_{n,\alpha}\hat{a}_{n+1,\alpha})+\sum_n Q\{(\hat{a}^+_{n,1}\hat{a}_{n,2}+\hat{a}^+_{n,2}\hat{a}_{n,1})+ \qquad (12\text{-}2)$$

$$+(\hat{a}^+_{n,2}\hat{a}_{n,3}+\hat{a}^+_{n,3}\hat{a}_{n,2})+(\hat{a}^+_{n,3}\hat{a}_{n+1,1}+\hat{a}^+_{n+1,1}\hat{a}_{n,3})\}+\sum_n Q'\{(\hat{a}^+_{n,1}\hat{a}^+_{n,2}-$$

$$-\hat{a}_{n,1}\hat{a}_{n,2})+(\hat{a}^+_{n,2}\hat{a}^+_{n,3}-\hat{a}_{n,2}\hat{a}_{n,3})+(\hat{a}^+_{n,3}\hat{a}^+_{n+1,1}-\hat{a}_{n,3}\hat{a}_{n+1,1})\}.$$

Here we used the following notations: $M = M_{n,\alpha;n+1,\alpha}$, $M' = M'_{n,\alpha;n+1,\alpha}$, $Q = M_{n,\alpha;n,\beta}$, $Q' = M'_{n,\alpha;n,\beta}$. We omitted also the indexes of the coefficients M, M', Q, Q'. This is possible due to the property of regularity of the model system (see Fig. 18).

Now, let us write the equations of motion for the six operators $\hat{a}_{n,1}$, $\hat{a}_{n,2}$, $\hat{a}_{n,3}$, $\hat{a}^+_{n,1}$, $\hat{a}^+_{n,2}$, $\hat{a}^+_{n,3}$ in the representation of Heisenberg

$$i\frac{d\hat{a}_{n,1}}{dt}=\Delta\hat{a}_{n,1}+\mathcal{M}(\hat{a}_{n+1,1}+\hat{a}_{n-1,1})+\mathcal{M}'(\hat{a}^+_{n+1,1}-\hat{a}^+_{n-1,1})+Q(\hat{a}_{n,2}+\hat{a}_{n-1,3})+Q'(\hat{a}^+_{n,2}-\hat{a}^+_{n-1,3}),$$

$$i\frac{d\hat{a}^+_{n,1}}{dt}=-\Delta\hat{a}^+_{n,1}-\mathcal{M}(\hat{a}^+_{n+1,1}+\hat{a}^+_{n-1,1})-\mathcal{M}'(\hat{a}_{n+1,1}-\hat{a}_{n-1,1})-Q(\hat{a}^+_{n,2}+\hat{a}^+_{n-1,3})-Q'(\hat{a}_{n,2}-\hat{a}_{n-1,3}),$$

$$i\frac{d\hat{a}_{n,2}}{dt}=\Delta\hat{a}_{n,2}+\mathcal{M}(\hat{a}_{n+1,2}+\hat{a}_{n-1,2})+\mathcal{M}'(\hat{a}^+_{n+1,2}-\hat{a}^+_{n-1,2})+Q(\hat{a}_{n,1}+\hat{a}_{n,3})+Q'(\hat{a}^+_{n,3}-\hat{a}^+_{n,1}),$$

$$i\frac{d\hat{a}^+_{n,2}}{dt}=-\Delta\hat{a}^+_{n,2}-\mathcal{M}(\hat{a}^+_{n+1,2}+\hat{a}^+_{n-1,2})-\mathcal{M}'(\hat{a}_{n+1,2}-\hat{a}_{n-1,2})-Q(\hat{a}^+_{n,1}+\hat{a}^+_{n,3})-Q(\hat{a}_{n,3}-\hat{a}_{n,1}), \qquad (12\text{-}3)$$

$$i\frac{d\hat{a}_{n,3}}{dt}=\Delta\hat{a}_{n,3}+\mathcal{M}(\hat{a}_{n+1,3}+\hat{a}_{n-1,3})+\mathcal{M}'(\hat{a}^+_{n+1,3}-\hat{a}^+_{n-1,3})+Q(\hat{a}_{n,2}+\hat{a}_{n+1,1})+Q'(\hat{a}^+_{n+1,1}-\hat{a}^+_{n,2}),$$

$$i\frac{d\hat{a}^+_{n,3}}{dt}=-\Delta\hat{a}^+_{n,3}-\mathcal{M}(\hat{a}^+_{n+1,3}+\hat{a}^+_{n-1,3})-\mathcal{M}'(\hat{a}_{n+1,3}-\hat{a}_{n-1,3})-Q(\hat{a}^+_{n,2}+\hat{a}^+_{n+1,1})-Q'(\hat{a}_{n+1,1}-\hat{a}_{n,2})$$

Taking into account the form of eqs. (12-3) let us introduce the six Green functions

$$G_{n-m;\alpha,1}=\langle\langle\hat{a}_{n,\alpha}|\hat{a}^+_{m,1}\rangle\rangle,\quad G'_{n-m;\alpha,1}=\langle\langle\hat{a}^+_{n,\alpha}|\hat{a}^+_{m,1}\rangle\rangle, \qquad (12\text{-}4)$$

$(\alpha = 1, 2, 3)$

and obtain the equations of motions for Fourier-components of the

Green functions

$$
\begin{aligned}
(E-T_k)G_{k;1,1} &= \frac{i}{2\pi} + iL_k G'_{k;1,1} + QG_{k;2,1} + Qe^{-ika}G_{k;3,1} + Q'G'_{k;2,1} - Q'e^{-ika}G'_{k;3,1},\\
(E+T_k)G'_{k;1,1} &= -iL_k G_{k;1,1} - QG'_{k;2,1} - Qe^{-ika}G'_{k;3,1} - Q'G_{k;2,1} + Q'e^{-ika}G_{k;3,1},\\
(E-T_k)G_{k;2,1} &= iL_k G'_{k;2,1} + QG_{k;1,1} + QG_{k;3,1} + Q'G'_{k;3,1} - Q'G'_{k;1,1},\\
(E+T_k)G'_{k;2,1} &= -iL_k G_{k;2,1} - QG'_{k;1,1} - QG'_{k;3,1} - Q'G_{k;3,1} + Q'G_{k;1,1},\\
(E-T_k)G_{k;3,1} &= iL_k G'_{k;3,1} + QG_{k;2,1} + Qe^{ika}G_{k;1,1} + Q'e^{ika}G'_{k;1,1} - Q'G'_{k;2,1},\\
(E+T_k)G'_{k;3,1} &= -iL_k G_{k;3,1} - QG'_{k;2,1} - Qe^{ika}G'_{k;1,1} - Q'e^{ika}G_{k;1,1} + Q'G_{k;2,1},
\end{aligned}
\qquad (12\text{-}5)
$$

where the values k, a, T_k and L_k are determined in the same way as in §10.

The condition of existence of non-trivial solution of the system of linear eqs. (12-5) gives us the dispersion equation determining the spectrum of elementary excitations

$$
\begin{vmatrix}
(E-T_k) & -iL_k & -Q & -Q' & -Qe^{-ika} & Q'e^{-ika}\\
iL_k & (E+T_k) & Q' & Q & -Q'e^{-ika} & Qe^{-ika}\\
-Q & Q' & (E-T_k) & -iL_k & -Q & -Q'\\
-Q' & Q & iL_k & (E+T_k) & Q' & Q\\
-Qe^{ika} & -Q'e^{ika} & -Q & Q' & (E-T_k) & -iL_k\\
Q'e^{ika} & Qe^{ika} & -Q' & Q & iL_k & (E+T_k)
\end{vmatrix} = 0. \qquad (12\text{-}6)
$$

At the beginning let us consider some particular solutions of eq. (12-6):

a) in the case when Q=Q'=0, that is when interactions between the third neighbors are very small and can be neglected we have

$$\{ E^2 - T_K^2 - L_K^2 \}^3 = 0. \tag{12-7}$$

Solving eq. (12-7) and omitting solutions with minus, which have not physical sense, we obtain

$$E_{1,2}(\kappa) = (T_K^2 + L_K^2)^{1/2}. \tag{12-8}$$

The result coincides with result (10-9) obtained for a molecular chain with one "molecule" in the lattice cell;

b) in the case when Q'=M'=0 known in physics as the Heitler-London approximation, eq. (12-7) takes the form

$$[(E - T_K)^3 - 2Q^3 \cos ka - 3Q^2 (E - T_K)] \cdot [(E + T_K)^3 + 2Q^3 \cos ka - 3Q^2 (E + T_K)] = 0 \tag{12-9}$$

Positive solutions of this equation are determined by the formulas

$$E_1(k) = T_k + 2Q\cos(ka/3),$$
$$E_{2,3}(k) = T_k + 2Q\cos[(ka \pm 2\pi)/3]; \qquad (12\text{-}10)$$

c) in the general case eq. (12-8) transforms to cubic equation in the unknown variable E^2. However, both the equation by itself and its solutions are rather bulky. So we shall give here only the results for the case k=0. In this case eq. (12-6) takes the form

$$[(E-T_0)^3 - 3(E-T_0)Q^2 - 2Q^3]\cdot[(E+T_0)^3 - 3(E+T_0)Q^2 + 2Q^3] -$$
$$-6Q'^2[(E-T_0)^2 - Q(E-T_0) - 2Q^2]\cdot[(E+T_0)^2 + Q(E+T_0) - 2Q^2] + \qquad (12\text{-}11)$$
$$+9Q'^4[(E-T_0) - 2Q]\cdot[(E+T_0) + 2Q] = 0,$$

where $T_0 = T_{k=0}$. After a number of simple transformations, eq. (12-11) can be led to the simpler form

$$(E-T_0-2Q)(E+T_0+2Q)[(E-T_0+Q)(E+T_0-Q) - 3Q'^2] = 0, \qquad (12\text{-}12)$$

whence it is not difficult to find positive solutions determining the unknown spectrum

$$E_1 = T_0 + 2Q,$$
$$E_{2,3} = [(T_0-Q)^2 + 3Q'^2]^{1/2}. \qquad (12\text{-}13)$$

A schematic picture of the energy spectrum calculated in all three cases considered above is shown in Fig. 19. For simplicity instead of the zones we show here only the lines corresponding to the zero value of the wave number k (k=0). The splitting of the spectral line, which corresponds to the excited state (1), is clearly visible in the figure.

Correlation of these results with the experimental data available was discussed in details in [31]. In particular, several experimental works confirming the existence of the splitting were presented there. So, we can state that the theory considered is in a good qualitative agreement with the experimental data.

§13. Neutron scattering by biopolymers.

The method of scattering of thermal (or slow) neutrons is widely used in physics for studying properties of solids. The neutron scattering experiment consists in the following. The bundle of collimated monochromatic neutrons is directed at the sample, and the energy spectrum of neutrons scattering in some direction is measured. Thus, the value measured is the differential section of scattering, which is denoted in general case by $d^2\sigma/dE_{\vec{p}\,'}\,d\Omega$ and determined as a ratio of the number of neutrons scattering in unit time and in unit space angle, to the falling stream.

Thermal neutrons (300° K) have the wavelength $\lambda=1{,}6$ Å. It means that their energy and wavelength are comparable with those of elementary excitations in regular crystal structures. That is why the method of neutron scattering is widely used for studying crystals. So, if we assume that in the first approximation biopolymers can be considered as quasi-one-dimensional crystals, it is then natural to imagine that the method of scattering of neutrons must be useful for studying biopolymers too.

In this paragraph we shall show how the correlation function calculated in the previous paragraphs, can be used for calculation of differential section of scattering of thermal neutrons. We shall mainly follow the scheme proposed by Van Hove [33]. To simplify calculations we shall return again to the simplest model of biopolymers, namely to one-dimensional chain of identical "molecules" (see Fig. 4), and try to imagine the sample, consisting of many chains of such a type, the chains being oriented along the z axis and weakly interacting with one another.

In the first approximation these interactions can be neglected and the model system can be described with a good accuracy by the hamiltonian

$$\hat{\mathcal{H}} = \sum_n \Delta \hat{P}^+_n \hat{P}_n + \sum_n \mathcal{M} (\hat{P}^+_n \hat{P}_{n+1} + \hat{P}^+_{n+1} \hat{P}_n) + \sum_n \mathcal{M}' (\hat{P}^+_n \hat{P}^+_{n+1} + \hat{P}_n \hat{P}_{n+1}), \tag{13-1}$$

where $\hat{P}^+_n$, $\hat{P}_n$ are Pauli operators; Δ, M, M′ are the coefficients determined as in §10; the index n numerates "molecules" of the chain, which are the scattering-centers.

Construction of scattering amplitude operator. In the theory of scattering, the physical value A named the length (or the amplitude) of scattering, is of great importance. It is a phenomenological characteristic of scattering, and it can be determined by the formula

$$\sigma = 4\pi |A|^2, \tag{13-2}$$

where is the total section of scattering by some fixed center.

In the general case the amplitude of scattering is a function of the vector $\vec{R}$ which determines coordinates of the scattering cent Let us assume that the operator $\hat{A}(\vec{R})$ corresponds to the physical the value $A(\vec{R})$, and write this operator in the representation of the secondary quantization in accordance with rules (5-18)

$$\hat{A}(\vec{R}) = \sum_{f,g} \langle f | \hat{A}(\vec{R}) | g \rangle \hat{P}^+_{\vec{R},f} \hat{P}_{\vec{R},g} \tag{13-3}$$

Taking into account that f and g take only the values 0 and 1 (see §10), we obtain

$$\hat{A}(\vec{R}) = \langle 0|\hat{A}(\vec{R})|0\rangle + \langle 1|\hat{A}(\vec{R})|0\rangle \hat{p}^{+}_{\vec{R},0}\hat{p}_{\vec{R},1} + \langle 0|\hat{A}(\vec{R})|1\rangle \hat{p}^{+}_{\vec{R},1}\hat{p}_{\vec{R},0} +$$

$$+ \langle 1|\hat{A}(\vec{R})|1\rangle \hat{p}^{+}_{\vec{R},1}\hat{p}_{\vec{R},1} = \alpha_{\vec{R}} + \alpha_{\downarrow\vec{R}}\hat{P}_{\vec{R}} + \alpha_{\uparrow\vec{R}}\hat{P}^{+}_{\vec{R}} + \beta_{\vec{R}}\hat{N}_{\vec{R}}, \tag{13-4}$$

where $\hat{P}^{+}_{\vec{R}}$, $\hat{P}_{\vec{R}}$ are Pauli-operators which are determined by the formulas

$$\hat{P}^{+}_{\vec{R}} = \hat{p}^{+}_{\vec{R},1}\hat{p}_{\vec{R},0}, \quad \hat{P}_{\vec{R}} = \hat{p}^{+}_{\vec{R},0}\hat{p}_{\vec{R},1}, \quad \hat{N}_{\vec{R}} = \hat{P}^{+}_{\vec{R}}\hat{P}_{\vec{R}},$$

$$\alpha_{\vec{R}} = \langle 0|\hat{A}(\vec{R})|0\rangle, \quad \alpha_{\downarrow\vec{R}} = \langle 0|\hat{A}(\vec{R})|1\rangle, \quad \alpha_{\uparrow\vec{R}} = \langle 1|\hat{A}(\vec{R})|0\rangle,$$

$$\beta_{\vec{R}} = \langle 1|\hat{A}(\vec{R})|1\rangle - \langle 0|\hat{A}(\vec{R})|0\rangle.$$

The physical sense of the coefficients of eq. (13-4) is rather simple: $\langle 0|\hat{A}(\vec{R})|0\rangle = \alpha_{\vec{R}}$ is the amplitude of scattering of thermal neutrons by the $\vec{R}$-th "molecule" which is in the ground state (0); $\langle 1|\hat{A}(\vec{R})|1\rangle$ is the amplitude of scattering by the $\vec{R}$-th "molecule" which is in the excited state (1); $\langle 0|\hat{A}(\vec{R})|1\rangle = \alpha_{\downarrow\vec{R}}$ is that of scattering by the $\vec{R}$-th "molecule" associated with the transition of the "molecule" from the excited state (1) to the ground state (0) (this is the reason why we used the notation with the arrow downward); $\langle 1|\hat{A}(\vec{R})|0\rangle = \alpha_{\uparrow\vec{R}}$ is the amplitude of scattering, associated with the transition of the $\vec{R}$-th "molecule" from the ground state (0) to the excited one (1).

Calculation of differential section of scattering. Let us assume that $\hat{V}$ is the operator of interaction of neutrons with the scattering sample. According to the theory of "pseudopotential" proposed by Fermi [34], there is a simple relation between the operators $\hat{V}$ and $\hat{A}$

$$\hat{V}(\vec{R}) = \frac{2\pi}{m_n} \sum_n \hat{A}(\vec{R}_n)\, \delta(\vec{R}-\vec{R}_n), \qquad (13\text{-}5)$$

where $\vec{R}$ and m_n are the coordinate and mass of neutron, respectively; $\vec{R}_n$ is the coordinate of a scattering center ($n=1,2,\dots N$). As it appears from the definition, the differential section of scattering must be proportional to the probability of scattering, and the probability in turn is proportional to the squared module of the corresponding matrix element of the operator $\hat{V}$. As a result we can obtain the following formula for the differential scattering section

$$\frac{d^2\sigma}{dE_{\vec{p}'}\,d\Omega} = \frac{1}{2\pi}\frac{p'}{p}\int_{-\infty}^{+\infty} e^{i(E_{\vec{p}'}-E_{\vec{p}})t} \sum_{\vec{R}_n,\vec{R}_{n'}} \langle \hat{A}^{+}(\vec{R}_n,t), \hat{A}(\vec{R}_{n'},t)\rangle e^{i(\vec{p}-\vec{p}')(\vec{R}_n-\vec{R}_{n'})}\, dt. \qquad (13\text{-}6)$$

Inserting (13-4) into (13-6) we find

$$\frac{d^2\sigma}{dE_{\vec{p}'}\,d\Omega} = \frac{1}{2\pi}\frac{p'}{p}\int_{-\infty}^{+\infty} dt\, e^{i(E_{\vec{p}'}-E_{\vec{p}})t} \sum_{n,n'} \{ \overline{\alpha^{*}_{n}\alpha_{n'}} + (\overline{\beta^{*}_{n}\alpha_{n'}} + \overline{\alpha^{*}_{n}\beta_{n'}})\langle \hat{N}(0)\rangle +$$

$$+ \overline{\alpha^{*}_{\downarrow n}\alpha_{\downarrow n'}} \langle \hat{P}^{+}_{n}(0)\, \hat{P}_{n'}(t)\rangle + \overline{\alpha^{*}_{\uparrow n}\alpha_{\downarrow n'}} \langle \hat{P}_{n}(0)\, \hat{P}_{n'}(t)\rangle + \qquad (13\text{-}7)$$

$$+ \overline{\alpha^*_{\downarrow n}\alpha_{\uparrow n'}}\langle \hat{P}^+_n(0)\hat{P}^+_{n'}(t)\rangle + \overline{\alpha^*_{\uparrow n}\alpha_{\uparrow n'}}\langle \hat{P}_n(0)\hat{P}^+_{n'}(t)\rangle\Big\} e^{-i(p_x - p'_x)(n-n')} .$$

Here $\vec{p}$ and $\vec{p}\,'$ are the impulses of incident and scattered neutrons; $p=|\vec{p}|$; $p'=|\vec{p}\,'|$; $(E_{\vec{p}\,'}-E_{\vec{p}})$ is the energy which the neutrons transfer to the scattering system; $\overline{\ldots}$ is the average over distribution of isotopes; $\langle\ldots\rangle$ is the statistical average over the hamiltonian $\mathcal{H}$.

The most interesting part of the differential section of scattering is the contribution given by, inelastic incoherent scattering

$$\frac{d^2\sigma^{incoh.inel.}}{dE_{\vec{p}\,'}d\Omega} = \frac{N}{8\pi^2}\frac{p'}{p}\int_{-\infty}^{+\infty} dt\, e^{i(E_{\vec{p}\,'}-E_{\vec{p}})t}\Big\{\sigma_{\downarrow\downarrow}\langle \hat{P}^+_n(0)\hat{P}_n(t)\rangle + \tag{13-8}$$

$$+\sigma_{\uparrow\downarrow}\langle \hat{P}_n(0)\hat{P}_n(t)\rangle + \sigma_{\downarrow\uparrow}\langle \hat{P}^+_n(0)\hat{P}^+_n(t)\rangle + \sigma_{\uparrow\uparrow}\langle \hat{P}_n(0)\hat{P}^+_n(t)\rangle\Big\},$$

where

$$\sigma_{\downarrow\downarrow} = 4\pi(\overline{\alpha^*_{\downarrow n}\alpha_{\downarrow n}} - \overline{\alpha^*_{\downarrow n}}\,\overline{\alpha_{\downarrow n}}), \qquad \sigma_{\downarrow\uparrow} = 4\pi(\overline{\alpha^*_{\downarrow n}\alpha_{\uparrow n}} - \overline{\alpha^*_{\downarrow n}}\,\overline{\alpha_{\uparrow n}}),$$

$$\sigma_{\uparrow\downarrow} = 4\pi(\overline{\alpha^*_{\uparrow n}\alpha_{\downarrow n}} - \overline{\alpha^*_{\uparrow n}}\,\overline{\alpha_{\downarrow n}}), \qquad \sigma_{\uparrow\uparrow} = 4\pi(\overline{\alpha^*_{\uparrow n}\alpha_{\uparrow n}} - \overline{\alpha^*_{\uparrow n}}\,\overline{\alpha_{\uparrow n}}).$$

Inserting the results of the calculation of correlation functions obtained in §10 into formula (13-8), we find

$$\frac{d^2\sigma^{incoh.inel.}}{dE_{\vec{p}\,'}d\Omega} = \frac{\varkappa}{4\pi N}\frac{p'}{p}\sum_k \Big\{\eta_k\,\delta(E_{\vec{p}\,'}-E_{\vec{p}}-E(k)) + (1-\eta_k)\,\delta(E_{\vec{p}\,'}-E_{\vec{p}}+E(k))\Big\}, \tag{13-9}$$

where $\varkappa = \sigma_{\uparrow\uparrow} = \sigma_{\downarrow\downarrow}$, $\eta_k = (1+\exp(E(k)/k_BT))^{-1}$; $E(k)$ is determined by eq. (10-9). The physical sense of the result obtained is rather simple. Delta-function in formula (13-9) guarantees the fulfillment of the law of conservation of energy. The first term in the braces describes the process of the absorption (annihilation) of elementary excitation with the energy $E(k)$ (Fig. 20-a), the second one describes the process of the creation of elementary excitation with the energy $E(k)$ (Fig. 20-b).

In conclusion, it is necessary to note that the method described above ,can be easily extended to other (more complicated models of biopolymers considered in this Part.

§14. Changes of exciton spectrum caused by small disturbances of the structure

Till now we discussed only the regular models of biopolymers. We considered chains with a strictly periodical sequence of "molecules"

A A A A. . .

A B A B A B. . .

and so on.

In this paragraph we shall illustrate the method by which the Green functions calculated above can be used for studying the effects of irregularity. We shall confine ourselves to consideration of one very simple example of the breaking of the regular structure when only one "molecule" of the regular chain

A A A A. . . ,

is substituted by the "molecule" of the type B (Fig. 21) An irregularity of this type is named a point impurity. The problem consists in finding the changes in the spectrum of elementary excitations caused by the substitution of one "molecule".

Let us write the hamiltonian describing the lowest excited states of the chain with one substituted "molecule"

$$\hat{\mathcal{H}} = \sum_n \Delta \hat{P}_n^+ \hat{P}_n + \sum_n \mathcal{M} (\hat{P}_n^+ \hat{P}_{n+1} + \hat{P}_{n+1}^+ \hat{P}_n) + \sum_n \mathcal{M}' (\hat{P}_n^+ \hat{P}_{n+1}^+ + \hat{P}_n \hat{P}_{n+1}) +$$
$$+ \delta \hat{P}_{n_0}^+ \hat{P}_{n_0} + m (\hat{P}_{n_0}^+ \hat{P}_{n_0-1} + \hat{P}_{n_0-1}^+ \hat{P}_{n_0} + \hat{P}_{n_0+1}^+ \hat{P}_{n_0} + \hat{P}_{n_0}^+ \hat{P}_{n_0+1}) + \qquad (14\text{-}1)$$

$$+m'(\hat{P}^{+}_{n_0}\hat{P}^{+}_{n_0+1}+\hat{P}^{+}_{n_0}\hat{P}^{+}_{n_0-1}+\hat{P}_{n_0}\hat{P}_{n_0+1}+\hat{P}_{n_0}\hat{P}_{n_0-1}).$$

Here $\hat{P}^{+}_n$, $\hat{P}_n$ are Pauli operators; the coefficients Δ, M and M' are determined as in §10; σ, m and m' are the changes of the coefficients due to the presence of the impurity.

Then we rewrite hamiltonian (14-1) in terms of Fermi-operators

$$\hat{\mathcal{H}}=\sum_n \Delta\hat{a}^{+}_n\hat{a}_n+\sum_n \mathcal{M}(\hat{a}^{+}_n\hat{a}_{n+1}+\hat{a}^{+}_{n+1}\hat{a}_n)+\sum_n \mathcal{M}'(\hat{a}^{+}_n\hat{a}^{+}_{n+1}-\hat{a}_n\hat{a}_{n+1})+$$
$$+\sigma\hat{a}^{+}_{n_0}\hat{a}_{n_0}+m(\hat{a}^{+}_{n_0-1}\hat{a}_{n_0}+\hat{a}^{+}_{n_0}a_{n_0-1}+\hat{a}^{+}_{n_0}\hat{a}_{n_0+1}+\hat{a}^{+}_{n_0+1}\hat{a}_{n_0})+ \qquad (14\text{-}2)$$
$$+m'(\hat{a}^{+}_{n_0}\hat{a}^{+}_{n_0+1}+\hat{a}^{+}_{n_0-1}\hat{a}^{+}_{n_0}-\hat{a}_{n_0}\hat{a}_{n_0+1}-\hat{a}_{n_0-1}\hat{a}_{n_0}).$$

To obtain the spectrum of elementary excitations of the chain with the impurity we shall follow the method worked out in the works of Lifshitz, Montroll, Potts, Dawber, Elliot and Maradudin [35-41]. The first stage of their method consists in obtaining the equations of motions for the spectral densities $\mathcal{J}^{(j)}(\omega)$. For this purpose let us construct the equations of motions for the operators $\hat{a}_\nu$, $\hat{a}^{+}_\nu$

$$i\frac{d\hat{a}_\nu}{dt}=\Delta\hat{a}_\nu+\mathcal{M}(\hat{a}_{\nu+1}+\hat{a}_{\nu-1})+\mathcal{M}'(\hat{a}^{+}_{\nu+1}-\hat{a}^{+}_{\nu-1})+\delta_{n_0,\nu}[\sigma\hat{a}_\nu+m(\hat{a}_{\nu+1}+\hat{a}_{\nu-1})+$$
$$+m'(\hat{a}^{+}_{\nu+1}-\hat{a}^{+}_{\nu-1})]+\delta_{n_0-1,\nu}(m\hat{a}_{\nu+1}+m'\hat{a}^{+}_{\nu+1})+\delta_{n_0+1,\nu}(m\hat{a}_{\nu-1}-m'\hat{a}^{+}_{\nu-1}),$$

$$i\frac{d\hat{a}^{+}_\nu}{dt}=-\Delta\hat{a}^{+}_\nu-\mathcal{M}(\hat{a}^{+}_{\nu+1}+\hat{a}^{+}_{\nu-1})-\mathcal{M}'(\hat{a}_{\nu+1}-\hat{a}_{\nu-1})-\delta_{n_0,\nu}[\sigma\hat{a}^{+}_\nu+ \qquad (14\text{-}3)$$
$$+m(\hat{a}^{+}_{\nu+1}+\hat{a}^{+}_{\nu-1})+m'(\hat{a}_{\nu+1}-\hat{a}_{\nu-1})]-\delta_{n_0-1,\nu}(m\hat{a}^{+}_{\nu+1}+m'\hat{a}_{\nu+1})-$$
$$-\delta_{n_0+1,\nu}(m\hat{a}^{+}_{\nu-1}-m'\hat{a}_{\nu-1}).$$

It is convenient to rewrite these equations in the more compact form

$$\sum_{\nu} (\varphi_{n,\nu} + \delta\varphi_{n,\nu})\hat{a}_{\nu} + (\psi_{n,\nu} + \delta\psi_{n,\nu})\hat{a}^{+}_{\nu} = 0,$$

$$\sum_{\nu} (\varphi^{*}_{n,\nu} + \delta\varphi_{n,\nu})\hat{a}^{+}_{\nu} + (\psi_{n,\nu} + \delta\psi_{n,\nu})\hat{a}_{\nu} = 0, \qquad (14\text{-}4)$$

where the following notations are used:

$$\varphi_{n,\nu} = -(i\frac{d}{dt} + \Delta)\delta_{n,\nu} - \mathcal{M}(\delta_{n+1,\nu} + \delta_{n-1,\nu}; \ \delta\varphi_{n,\nu} = -\delta_{n_0,n}\{\theta\delta_{n,\nu} + m(\delta_{n+1,\nu} + \delta_{n-1,\nu})\} - m\,\delta_{n_0-1,n}\,\delta_{n+1,\nu} - m\,\delta_{n_0+1,n}\,\delta_{n-1,\nu}; \ \psi_{n,\nu} = \mathcal{M}'(\delta_{n+1,\nu} - \delta_{n-1,\nu});$$

$$\delta\psi_{n,\nu} = m'\delta_{n_0,n}(\delta_{n+1,\nu} - \delta_{n-1,\nu}) + m'\delta_{n_0-1,n}\,\delta_{n+1,\nu} - m'\delta_{n_0+1,n}\,\delta_{n-1,\nu}.$$

In accordance with the form of eqs. (14-4), let us introduce four correlation functions: $\langle\hat{a}_n(t)\hat{a}^{+}_m(0)\rangle$, $\langle\hat{a}_n(t)\hat{a}_m(0)\rangle$, $\langle\hat{a}^{+}_n(t)\hat{a}^{+}_m(0)\rangle$, $\langle\hat{a}^{+}_n(t)\hat{a}_m(0)\rangle$, and corresponding spectral densities $\mathcal{J}^{(j)}_{n,m}(\omega)$ $(j=1,2,3,4)$

$$\langle\hat{a}_n(t)\hat{a}^{+}_m(0)\rangle = \int_{-\infty}^{+\infty} \mathcal{J}^{(1)}_{n,m}(\omega)e^{i\omega t}d\omega,$$

$$\langle\hat{a}_n(t)\hat{a}_m(0)\rangle = \int_{-\infty}^{+\infty} \mathcal{J}^{(2)}_{n,m}(\omega)e^{i\omega t}d\omega,$$

$$\langle\hat{a}^{+}_n(t)\hat{a}^{+}_m(0)\rangle = \int_{-\infty}^{+\infty} \mathcal{J}^{(3)}_{n,m}(\omega)e^{i\omega t}d\omega, \qquad (14\text{-}5)$$

$$\langle\hat{a}^{+}_n(t)\hat{a}_m(0)\rangle = \int_{-\infty}^{+\infty} \mathcal{J}^{(4)}_{n,m}(\omega)e^{i\omega t}d\omega.$$

Using the results obtained in §8, it is not difficult to obtain the equations of motion for correlation functions (14-5). Then we can expand these equations in Fourier integral taken over time. As a result we obtain the equations of motions for the spectral densities $\mathcal{J}^{(j)}_{\nu,m}(\omega)$

$$\sum_{\nu} (\Phi_{n,\nu}+\delta\Phi_{n,\nu}) J^{(4)}_{\nu,m}(\omega) - (\Psi_{n,\nu}+\delta\Psi_{n,\nu}) J^{(2)}_{\nu,m}(\omega) = 0,$$

$$\sum_{\nu} (\Psi_{n,\nu}+\delta\Psi_{n,\nu}) J^{(4)}_{\nu,m}(\omega) + (\Phi'_{n,\nu}-\delta\Phi_{n,\nu}) J^{(2)}_{\nu,m}(\omega) = 0,$$

$$\sum_{\nu} (\Phi_{n,\nu}+\delta\Phi_{n,\nu}) J^{(3)}_{\nu,m}(\omega) - (\Psi_{n,\nu}+\delta\Psi_{n,\nu}) J^{(1)}_{\nu,m}(\omega) = 0, \qquad (14\text{-}6)$$

$$\sum_{\nu} (\Psi_{n,\nu}+\delta\Psi_{n,\nu}) J^{(3)}_{\nu,m}(\omega) - (\Phi'_{n,\nu}-\delta\Phi_{n,\nu}) J^{(1)}_{\nu,m}(\omega) = 0,$$

where the new notations are introduced

$$\Phi_{n,\nu} = (\omega-\Delta)\delta_{n,\nu} - M(\delta_{n+1,\nu}+\delta_{n-1,\nu}); \quad \Phi'_{n,\nu} = (\omega+\Delta)\delta_{n,\nu} + M(\delta_{n+1,\nu}+\delta_{n-1,\nu}).$$

Let us rewrite eqs. (14-6) in more compact (matrix) form

$$(\mathbf{L} + \delta\mathbf{L})\cdot\mathbf{J} = \mathbf{0}, \qquad (14\text{-}7)$$

where the composite four-block matrices

$$\mathbf{L} = \begin{pmatrix} \Phi & -\Psi \\ \Psi & \Phi' \end{pmatrix}, \quad \delta\mathbf{L} = \begin{pmatrix} \delta\Phi & -\delta\Psi \\ \delta\Psi & -\delta\Phi \end{pmatrix}, \quad \mathbf{J} = \begin{pmatrix} J^{(4)} & J^{(3)} \\ J^{(2)} & J^{(1)} \end{pmatrix}$$

consist of the matrices: $\Phi = \|\Phi_{n,\nu}\|$, $\Phi' = \|\Phi'_{n,\nu}\|$,

$\Psi = \|\Psi_{n,\nu}\|$, $\delta\Phi = \|\delta\Phi_{n,\nu}\|$, $\delta\Psi = \|\delta\Psi_{n,\nu}\|$, $J^{(j)} = \|J^{(j)}_{n,\nu}(\omega)\|$, $j = 1,2,3,4$.

For ideal chain: $\delta\mathbf{L} = \mathbf{0}$.

Matrix equation (14-7) can be transformed into equivalent form

$$\mathbf{L}\cdot(\mathbf{I}+\mathbf{L}^{-1}\cdot\delta\mathbf{L})\cdot\mathbf{J}=0, \qquad (14-8)$$

where $\mathbf{L}^{-1}$ is the inverse matrix of $\mathbf{L}$, $\mathbf{I}=\begin{pmatrix} I & 0 \\ 0 & I \end{pmatrix}$ is the composite matrix, consisting of the zero ($0=\|0\|$) and the identical ($I=\|\delta_{n,\nu}\|$) matrices. Of course, the condition

$$\det \mathbf{L} \neq 0$$

is assumed to be fulfilled.

The next stage of the method consists in finding the matrix $\mathbf{L}^{-1}$ appearing in formula (14-8). To this purpose let us use the results obtained in §10 for the chain without the impurity. We shall name this chain "the ideal chain". Let us reproduce, for example, the equations of motion for Fourier-components in time for four Green functions: $G^{(1)}_{n-m}(E)=\langle\langle\hat{a}_n|\hat{a}^+_m\rangle\rangle_E$, $G^{(2)}_{n-m}(E)=\langle\langle\hat{a}^+_n|\hat{a}^+_m\rangle\rangle_E$, $G^{(3)}_{n-m}(E)=\langle\langle\hat{a}_n|\hat{a}_m\rangle\rangle$, $G^{(4)}_{n-m}(E)=\langle\langle\hat{a}^+_n|\hat{a}_m\rangle\rangle$

$$(E-\Delta)G^{(1)}_{n-m}(E)=\frac{i\delta(n-m)}{2\pi}+\mathcal{M}\{G^{(1)}_{n+1-m}(E)+G^{(1)}_{n-1-m}(E)\}+\mathcal{M}'\{G^{(2)}_{n+1-m}(E)-G^{(2)}_{n-1-m}(E)\},$$

$$(E+\Delta)G^{(2)}_{n-m}(E)=-\mathcal{M}\{G^{(2)}_{n+1-m}(E)+G^{(2)}_{n-1-m}(E)\}-\mathcal{M}'\{G^{(1)}_{n+1-m}(E)-G^{(1)}_{n-1-m}(E)\},$$

$$(E-\Delta)G^{(3)}_{n-m}(E)=\mathcal{M}\{G^{(3)}_{n+1-m}(E)+G^{(3)}_{n-1-m}(E)\}+\mathcal{M}'\{G^{(4)}_{n+1-m}(E)-G^{(4)}_{n-1-m}(E)\},$$

$$(14-9)$$

$$(E+\Delta)G^{(4)}_{n-m}(E)=\frac{i\delta(n-m)}{2\pi}-\mathcal{M}\{G^{(4)}_{n+1-m}(E)+G^{(4)}_{n-1-m}(E)\}-\mathcal{M}'\{G^{(3)}_{n+1-m}(E)-G^{(3)}_{n-1-m}(E)\}.$$

Let us rewrite these equations in a more compact (matrix) form

$$\mathbf{L}\ \mathbf{G} = \frac{i}{2\pi}\ \mathbf{I}, \qquad (14\text{-}10)$$

where $\mathbf{G}=\begin{pmatrix} G^{(1)} & G^{(3)} \\ G^{(2)} & G^{(4)} \end{pmatrix}$ is the composite matrix consisting of the matrices whose elements are the Fourier-components of the Green functions of the ideal chain

$$G^{(1)} = \| G^{(1)}_{n,m}(E) \|,\ G^{(2)} = \| G^{(2)}_{n,m}(E) \|,\ G^{(3)} = \| G^{(3)}_{n,m}(E) \|,\ G^{(4)} = \| G^{(4)}_{n,m}(E) \| .$$

Multiplying (14-10) by $\mathbf{L}^{-1}$ (from the left side) and taking into account that $\mathbf{L}^{-1} \cdot \mathbf{L} = \mathbf{I}$, we find the unknown matrix $\mathbf{L}^{-1}$

$$\mathbf{L}^{-1} = \frac{2\pi}{i}\ \mathbf{G}. \qquad (14\text{-}11)$$

Then, after inserting (14-11) in eq. (14-8)

$$\mathbf{L}\cdot(\ \mathbf{I} + \frac{2\pi}{i}\ \mathbf{G}\cdot\delta\mathbf{L})\cdot\mathbf{J} = \mathbf{0} \qquad (14\text{-}12)$$

and using the condition of the existence of nontrivial solution of eq. (14-12) we obtain the formula

$$\det\ (\ \mathbf{I} + \frac{2\pi}{i}\ \mathbf{G}\cdot\delta\mathbf{L}) = \mathbf{0}, \qquad (14\text{-}13)$$

which determines the spectrum of elementary excitations in the chain with the point impurity.

If we insert now the explicit form of the Green function of the ideal chain into (14-13) and leave the terms which are linear according to σ, m and m', we obtain finally

$$1 = \frac{2}{N} \sum_{\kappa} \frac{T_{\kappa}(\delta + 4m\cos\kappa a) + 4L_{\kappa} m' \sin\kappa a}{E^{2} - E^{2}(\kappa)} \qquad (14-14)$$

where k, T_k, L_k and E(k) are determined in the same way as in §10. Note that the condition of small amplitude of perturbation associated with the impurity is used only at this stage of the scheme of the calculation.

In the particular case when m=m′=0 (that is in the case when the "molecule" of the impurity differs from other "molecules" of the chain only by the value of the energy of excitation), eq. (14-14) takes a simpler form

$$1 = \frac{2\delta}{N} \sum_{\kappa} \frac{T_{\kappa}}{E^{2} - E^{2}(\kappa)} . \qquad (14-15)$$

The structure of eq. (14-15) is equivalent to that obtained by Maradudin [40] when studying the spectra of crystals with isotopic impurity. Let us reproduce here the main qualitative results obtained by him. A schematic picture of the graphical solutions of eq. (14-15) is shown in Fig. 22. The points of intersection of the curves with the horizontal line passing through the point 1 on the ordinate axis, determine the frequencies* of the spectrum of the chain with the impurity.

---

* See the footnote in the page [illegible]

Additional vertical lines are shown: they cross the abscissa axis at points which correspond to the values of the squares of the frequencies of the unperturbed spectrum (that is of the spectrum of the ideal chain). The ensemble of these lines forms a zone (or a band).

Let us consider at first the case $\delta<0$. The fulfillment of this condition means that the energy of the first excited state (1) of the impurity "molecule" is smaller than that of the basic "molecules" of the chain.

As follows from Fig. 22a the crossover points are on the left of the vertical lines. It means that the presence of the impurity molecule leads to a small displacement of the frequencies to the left, that is the frequencies slightly decrease in comparison with the analogous frequencies of the ideal chain.

In the opposite case ($\delta>0$), when the energy of the first excited state (1) of the impurity "molecule" is larger than that of the basic "molecules" of the chain, a small displacement of the frequencies to the right are observed (Fig. 22b), the displacement of the highest mode can be rather large and to get out of the bound of the zone.

It is not difficult now to extend these results to the more general case when the chain contains not only one but several impurity "molecules". Indeed, let us assume that the density of the impurity "molecules" is rather small. Then we can suggest that the distances between them are rather large, and the correlations between the impurity "molecules" can be neglected. So, one can expect that these "molecules" will give additive contribution to the spectrum, and hence the qualitative results on the spectral changes due to the impurity will be the same. The only difference is the appearance of an additional factor which is proportional to the number of the impurity molecules.

§15. Striction effects.

In this paragraph we shall take one more step towards the improving of the physical models of biopolymers. Till now we suggested that the "molecules" forming the chain are rigidly fixed at the lattice points. Only the internal degrees of freedom of the "molecules" described, for example, by the energy levels (see Fig. 12), were taken into account. Now we remove this limitation and assume that the "molecules" can oscillate near the lattice points. The problem consists in finding changes of both the spectrum of elementary excitations (or excitons) and the correlation functions, which are caused by the thermal oscillations of the "molecules". It is interesting also to investigate the inverse influence of the excitons on the oscillations of the "molecules".

Let us begin with the construction of the model hamiltonian. We can expect that it consists of three main terms

$$\hat{\mathcal{H}} = \hat{\mathcal{H}}_1 + \hat{\mathcal{H}}_2 + \hat{\mathcal{H}}_{int} \qquad (15\text{-}1)$$

where $\hat{\mathcal{H}}_1$ is the hamiltonian describing oscillations of the "molecules" in the chain in the absence of any excitations; $\hat{\mathcal{H}}_2$ is the hamiltonian describing excited states of the chain with hard fixed "molecules"; $\hat{\mathcal{H}}_{int.}$ describes interaction between excitons and thermal oscillations of the "molecules". For simplicity let us consider here a chain of identical molecules and suppose that only longitudinal oscillations are allowed. Then the hamiltonian $\hat{\mathcal{H}}_1$ takes the form

$$\hat{\mathcal{H}}_1 = \sum_n \frac{\mathcal{P}_n^2}{2m} + \frac{1}{2}\sum_n m\omega_0^2 (u_{n+1} - u_n)^2, \tag{15-2}$$

where m is the mass of the "molecules", ω_0 is the frequency of the oscillations, $\mathcal{P}_n$ is the impulse of the n-th "molecule", u_n is the displacement of the n-th "molecule".

To find the explicit form of the hamiltonian $\hat{\mathcal{H}}_2$ we can use formula (10-1) or the equivalent one (10-20)

$$\hat{\mathcal{H}}_2 = \sum_n (E_1 - E_0 + 2D)\hat{a}_n^+ \hat{a}_n + \sum_n \mathcal{M}(\hat{a}_n^+ \hat{a}_{n+1} + \hat{a}_{n+1}^+ \hat{a}_n) + \sum_n \mathcal{M}'(\hat{a}_n^+ \hat{a}_{n+1}^+ - \hat{a}_n \hat{a}_{n+1}), \tag{15-3}$$

where E_1 and E_0 are the energies of the excited and the ground states of one isolated "molecule"; the coefficients D, M and M′ are matrix elements of the operator of interaction between the "molecules" $\hat{V}_{n,n+1}$; $\hat{a}_n^+$, $\hat{a}_n$ are Fermi operators.

It remains only to find the explicit form of the term $\hat{\mathcal{H}}_{int.}$ To this purpose let us use the traditional technique based on a simple assumption about the behavior of coefficients D, M and M′ of the hamiltonian $\hat{\mathcal{H}}_2$. It is assumed that these coefficients increase or decrease as the distances between the "molecules" are changed. It means that the coefficients are to be functions of the distances. Then we can write the sum of two the terms $\hat{\mathcal{H}}_1$ and $\hat{\mathcal{H}}_2$) in the form

$$\hat{\mathcal{H}}_2 + \hat{\mathcal{H}}_{int.} = \sum_n (E_1 - E_0)\hat{a}_n^+ \hat{a}_n + \sum_n D(z_{n+1} - z_n)(\hat{a}_n^+ \hat{a}_n + \hat{a}_{n+1}^+ \hat{a}_{n+1}) + \tag{15-4}$$

$$+\sum_n \mathcal{M}(z_{n+1}-z_n)(\hat{a}_n^+\hat{a}_{n+1}+\hat{a}_{n+1}^+\hat{a}_n)+\sum_n \mathcal{M}'(z_{n+1}-z_n)(\hat{a}_n^+\hat{a}_{n+1}^+-\hat{a}_n\hat{a}_{n+1}).$$

Taking into account that the displacements of the "molecules" from the lattice points are small, let us expand the coefficients D, M and M′ in a power series of $(u_{n+1} - u_n)$

$$\mathcal{D}(z_{n+1}-z_n)=\mathcal{D}(a)\left\{1+\eta\frac{u_{n+1}-u_n}{a}+\dots\right\},$$

$$\mathcal{M}(z_{n+1}-z_n)=\mathcal{M}(a)\left\{1+\xi\frac{u_{n+1}-u_n}{a}+\dots\right\}, \qquad (15-5)$$

$$\mathcal{M}'(z_{n+1}-z_n)=\mathcal{M}'(a)\left\{1+\zeta\frac{u_{n+1}-u_n}{a}+\dots\right\},$$

where a is the lattice constant; η, ξ and ζ are determined by the formulas

$$\eta=[\partial\ln\mathcal{D}(z_{n+1}-z_n)/\partial\ln(z_{n+1}-z_n)]_{z_{n+1}-z_n=a},$$

$$\xi=[\partial\ln\mathcal{M}(z_{n+1}-z_n)/\partial\ln(z_{n+1}-z_n)]_{z_{n+1}-z_n=a}, \qquad (15-6)$$

$$\zeta=[\partial\ln\mathcal{M}'(z_{n+1}-z_n)/\partial\ln(z_{n+1}-z_n)]_{z_{n+1}-z_n=a}.$$

If we limit ourselves to the linear approximation in formulas (15-5) and put together all three terms of the operator $\hat{\mathcal{H}}$, we obtain the unknown model hamiltonian

$$\hat{\mathcal{H}}=\sum_n\frac{\mathcal{P}_n^2}{2m}+\frac{1}{2}\sum_n m\omega_0^2(u_{n+1}-u_n)^2+\sum_n(E_1-E_0)\hat{a}_n^+\hat{a}_n+$$

$$+\sum_n\mathcal{D}(a)\left(1+\eta\frac{u_{n+1}-u_n}{a}\right)(\hat{a}_n^+\hat{a}_n+\hat{a}_{n+1}^+\hat{a}_{n+1})+ \qquad (15-7)$$

$$+\sum_n\mathcal{M}(a)\left(1+\xi\frac{u_{n+1}-u_n}{a}\right)(\hat{a}_n^+\hat{a}_{n+1}-\hat{a}_{n+1}^+\hat{a}_n)+$$

$$+\sum_n \mathcal{M}'(a)\left(1+\mathcal{E}\frac{u_{n+1}-u_n}{a}\right)(\hat{a}^+_n\hat{a}^+_{n+1}-\hat{a}_n\hat{a}_{n+1}).$$

To simplify the solution of the problem (15-7), it is desirable to remove the cross terms in the hamiltonian $\hat{\mathcal{H}}$, which connect the Fermi-operators $\hat{a}^+_n$, $\hat{a}_n$ with the displacements of the "molecules" u_n. We can achieve this if we move the "start point" of the motions of the "molecules" by A_n

$$A_n=\frac{1}{m\omega_0^2 a}\sum_{\nu=1}^{n-1}\{\mathcal{D}\eta(\hat{a}^+_\nu\hat{a}_\nu+\hat{a}^+_{\nu+1}\hat{a}_{\nu+1})+\mathcal{M}\xi(\hat{a}^+_\nu\hat{a}_{\nu+1}+\hat{a}^+_{\nu+1}\hat{a}_\nu)+ \tag{15-8}$$

$$+\mathcal{M}'\mathcal{E}(\hat{a}^+_\nu\hat{a}^+_{\nu+1}-\hat{a}_\nu\hat{a}_{\nu+1})\}.$$

As a result the hamiltonian $\hat{\mathcal{H}}$ takes the form

$$\hat{\mathcal{H}}=\sum_n\frac{\mathcal{P}_n^2}{2m}+\frac{1}{2}\sum_n m\omega_0^2(\tilde{u}_{n+1}-\tilde{u}_n)^2+\sum_n\tilde{\Delta}\hat{a}^+_n\hat{a}_n+$$

$$+\sum_n\tilde{\mathcal{M}}(\hat{a}^+_n\hat{a}_{n+1}+\hat{a}^+_{n+1}\hat{a}_n)+\sum_n\tilde{\mathcal{M}}'(\hat{a}^+_n\hat{a}^+_{n+1}-\hat{a}_n\hat{a}_{n+1})+ \tag{15-9}$$

$$+\sum_n h\hat{N}_n\hat{N}_{n+1}+\text{Const.},$$

where

$$\tilde{u}_n=u_n+A_n,\quad \hat{N}_n=\hat{a}^+_n\hat{a}_n,\quad \tilde{\Delta}=E_1-E_0+2\tilde{\mathcal{D}},$$

$$\tilde{\mathcal{D}}=\mathcal{D}(a)-\frac{1}{2m\omega_0^2a^2}\left[\mathcal{D}(a)\eta^2+\mathcal{M}(a)\xi^2-\mathcal{M}'(a)\mathcal{E}^2\right],$$

$$\tilde{\mathcal{M}}=\mathcal{M}(a)\left[1-\eta\xi\frac{\mathcal{D}(a)}{m\omega_0^2a^2}\right],\quad \tilde{\mathcal{M}}'=\mathcal{M}'(a)\left[1-\eta\mathcal{E}\frac{\mathcal{D}(a)}{m\omega_0^2a^2}\right],$$

$$h=-\frac{1}{m\omega_0^2a^2}\left[\mathcal{D}^2(a)\eta^2-\mathcal{M}^2(a)\xi^2-\mathcal{M}'^2(a)\mathcal{E}^2\right].$$

The result (15-9) can be interpreted in the following way. The first two sums in the hamiltonian $\hat{\mathcal{H}}$ describe oscillations of the "molecules" having a frequency ω_0 which is the same as in $\hat{\mathcal{H}}_1$.

However, these oscillations are near the new lattice points which are moved away by the value A_n. The other terms of the hamiltonian H describe the excited states of the chain with the rigidly fixed "molecules". Comparing the latter terms with $\hat{\mathcal{H}}_2$ one can note that the influence of the phonon subsystem leads to two main results: (1) the renormalization of the coefficients Δ, M, M′ and (2) the appearance of the quartic term $\sum_n h\hat{N}_n\hat{N}_{n+1}$ which describes interaction between excitations. If we assume that $\sum_n h\hat{N}_n\hat{N}_{n+1} \cong 2\sum_n h\langle N\rangle \hat{N}_n$, the influence of the phonon subsystem leads only to the renormalization of the coefficients Δ, M and M′. Hence, it appears that both the correlation functions and the spectrum of elementary excitations have the same form as in §10, but with the renormalized coefficients Δ, M and M′.

An important consequence of the existence of the interaction between the exciton and phonon subsystems is the change of the average values of the distances between the "molecules" of the chain

$$\langle z_{n+1} - z_n \rangle = a - \langle A_{n+1} - A_n \rangle = a - \frac{1}{m\omega_0^2 a}\{2\mathcal{D}(a)\eta\langle \hat{a}_n^+ \hat{a}_n\rangle + \mathcal{M}(a)\xi\langle \hat{a}_n^+ \hat{a}_{n+1} + \hat{a}_{n+1}^+ \hat{a}_n\rangle + \mathcal{M}'(a)\xi\langle \hat{a}_n^+ \hat{a}_{n+1}^+ - \hat{a}_n \hat{a}_{n+1}\rangle\}. \quad (15\text{-}10)$$

In the approximation when M=M′=0 or $\xi = \xi = 0$ formula (15-10) takes the simple form

$$\langle z_{n+1} - z_n \rangle - a = -\frac{2\mathcal{D}(a)\eta\langle \hat{N}\rangle}{m a \omega_0^2} < 0, \quad (15\text{-}11)$$

It results from this that the average values of the distances between the "molecules" of the chain decrease, and the total length of the chain (L) is also decreased. This effect is named the "striction" [42] and we used this word in the title of this paragraph.

In this approximation it is easy to show that the coefficient of the quartic term is equal to

$$h = -\frac{\mathcal{D}^2(a)\eta^2}{ma^2\omega_0^2} < 0. \qquad (15\text{-}12)$$

It follows from this that the influence of the oscillations of the "molecules" on the exciton subsystem leads to the appearance of the weak attraction between the excitations which were earlier seen as an ideal Fermi-gas.

PART 4. ELEMENTS OF NONLINEAR WAVE THEORY.

Nonlinear wave theory is the field of contemporary theoretical physics dealing with the study of propagation of waves in nonlinear media. Mathematical basis of the theory is now in the process of active development. The basis is formed by several nonlinear wave equations and the methods of finding, analysis and interpretation of their solutions.

Some of the solutions are known as solitons. They play an important role in applications, and the nonlinear wave theory is called often the theory of solitons. In this Part some elements of the theory which are necessary to understand applications are considered. The reader who wishes to find a more detailed description of the nonlinear theory can be refered to the reviews of Scott, Chu, McLaughlin [43] and Makhankov, Fedyanin [44], to the collections of articles [45,46] and to the books of Calogero, Degasperis [47], Whitham [48] and others.

§16. $\hat{a} \rightarrow \Psi$ transformation

Before starting I would like to call your attention to one important feature for which the material of this Part differs from that of others. In the previous Parts (the second and the third ones) the subject of the investigation was quantum systems, and different quantum operators were used to describe the systems. Among the operators the hamiltonian $\hat{\mathcal{H}}$ written in the secondary quantum representation played an especially important role. The dynamics of the systems were described by the equations of motions for the quantum operators of creation and annihilation $\hat{a}^+$, $\hat{a}$ in the representation of Heisenberg.

In this and in the next Parts we shall consider classical systems. As we mentioned in §4, the state and the evolution of these systems can be described by two formalisms: the formalism of Lagrange and that of Hamilton. Both formalisms deal with usual classical functions which are determined in the space of coordinates and time. Among them the functions of Lagrange and Hamilton play the most important role. Using them we can easily construct equations of motion determining the dynamics of classical systems.

Can we find the relation between the equations of motions for quantum operators $\hat{a}^{+}$, $\hat{a}$ and the equations of motion for usual classical functions? Indeed, this relation exists. In this paragraph one of the methods of transformation of the equations of motions for Fermi-operators $\hat{a}^{+}$, $\hat{a}$ to those for functions Ψ^{*}, Ψ which are the amplitudes of some classical field, is described. For short we name this transformation the $\hat{a} \to \Psi$ transformation. For the first time it was derived in the work [49] where the authors studied the propagation of the impulse of excitation along one-dimensional molecular chain. Later this method was successfully used in many other problems. Propagation of nonlinear waves in quasi-one-dimensional magnets is one of them [50-51]. Another is the nonlinear dynamics of polyacetylene type systems [52]. One further example is the propagation of nonlinear waves in alpha-helical proteins [53-54].

The $a \to \Psi$ transformation can be considered as an analog of the temporal perturbation theory which is widely used in quantum mechanics. Let us recall the main points of the theory.

So, it is supposed that the hamiltonian of the dynamical system considered can be divided into two parts

$$\hat{\mathcal{H}} = \hat{\mathcal{H}}^{(0)} + \hat{V}(t). \qquad (16\text{-}1)$$

Here $\hat{V}(t)$ is a small perturbation, $\hat{\mathcal{H}}^{(0)}$ is unperturbed part of the hamiltonian, about which we know that: (1) it does not depend on time t; (2) its eigenfunctions $\varphi_f^{(0)}$ and eigenvalues $E_f^{(0)}$ are known, that is the exact solutions of the stationary equation of Schrödinger

$$\hat{\mathcal{H}}^{(0)} \varphi_f^{(0)} = E_f^{(0)} \varphi_f^{(0)}. \qquad (16\text{-}2)$$

are known. The problem consists in finding approximate solution of the wave equation of Schrödinger

$$i \frac{\partial \Psi}{\partial t} = (\hat{\mathcal{H}}^{(0)} + \hat{V}(t)) \Psi . \qquad (16\text{-}3)$$

Let us find the solution in the form of a sum of terms

$$\Psi = \sum_f a_f(t)\, e^{-iE_f^{(0)}} \varphi_f^{(0)}, \qquad (16\text{-}4)$$

where $a_f(t)$ are the coefficients depending on time t. Inserting (16-4) into (16-3), multiplying the left and the right of the equation obtained by $e^{-E_g^{(0)}t} \varphi_g^{(0)*}$, and integrating the result we have

$$i \frac{da_g}{dt} = \sum_f V_{gf}(t) a_f , \qquad (16\text{-}5)$$

where $V_{gf}(t) = e^{i(E_g^{(0)} - E_f^{(0)})} \int \varphi_g^{(0)*} \hat{V} \varphi_f^{(0)} dq.$

Let us expand the coefficients a_f

$$a_f = a_f^{(0)} + a_f^{(1)} + a_f^{(2)} + \cdots, \qquad (16\text{-}6)$$

where the term $a_f^{(1)}$ is of the same (i.e. the first) order as the perturbation $\hat{V}(t)$; the term $a_f^{(2)}$ is the second-order correction and so on. Inserting (16-6) into (16-5) and keeping only the first-order terms, we obtain

$$i \frac{da_g^{(1)}}{dt} = \sum_f V_{gf}(t)\, a_f^{(0)}. \qquad (16\text{-}7)$$

If we use the eigenfunction of the problem (16-2) $\varphi_{f'}^{(0)}$ as unperturbed wave function $\Psi^{(0)}$

$$\Psi^{(0)} \equiv \varphi_{f'}^{(0)} \qquad (16\text{-}8)$$

the coefficient $a_f^{(0)}$ in formula (16-7) becomes equal to

$$a_f^{(0)} = \begin{cases} 1, & \text{if } f = f' \\ 0, & \text{if } f = f'. \end{cases} \qquad (16\text{-}9)$$

Inserting (16-9) into (16-7), we find

$$i \frac{da_g^{(1)}}{dt} = V_{gf'}(t). \qquad (16\text{-}10)$$

Let us introduce an additional index for the coefficients a_f

$$a_f \rightarrow a_{ff'}, \qquad (16\text{-}11)$$

to indicate for which of the unperturbed functions $\varphi_f^{(0)}$ the correction is calculated. Then, instead of (16-4) we obtain

$$\Psi_{f'} = \sum_f a_{ff'}(t)\, e^{-iE_f^{(0)}t}\, \varphi_f^{(0)}, \qquad (16\text{-}12)$$

where the coefficients $a_{ff'}(t)$ obtained from (16-10) are equal to

$$a_{ff'}^{(1)} = -i \int V_{ff'}(t)\, dt. \qquad (16\text{-}13)$$

The result (16-12) can be generalized and written in a more compact form

$$\Psi(t) = \hat{U}(t)\, \Psi(0), \qquad (16\text{-}14)$$

where $\Psi(0)$ is the wave function of the state of the system at time $t=0$; $\Psi(t)$ is that at time t; $\hat{U}(t)$ is the operator of evolution, which transforms $\Psi(0)$ to $\Psi(t)$. It is obvious that the operator $\hat{U}$ should be hermitian ($\hat{U}^+ = \hat{U}$) and unitary ($\hat{U}^+\hat{U} = \hat{U}\hat{U}^+ = 1$).

Coming back to the theme of this paragraph, let us try to apply the result (16-14) to the model systems determined in the secondary quantum representation. Assume that at time $t=0$ they are in the unexcited state

$$\Psi(0) = |N_1 = 0, \ldots N_f = 0, \ldots \rangle. \qquad (16\text{-}15)$$

This state is usually named the vacuum state and denoted by symbol $|0\rangle$. The problem consists in constructing the function $\Psi(t)$ which determines the state of the model system at time t. If we use formula (16-14) the problem considered can be reformulated in another way, namely the problem comes then to the one of construction of the operator $\hat{U}$.

It is natural to suggest that in the considered problems where all events occur in the space of operators of the secondary quantization (or more precisely to say in the space of Fermi-operators $\hat{a}^+$, $\hat{a}$), the operator $\hat{U}$ also must be a function of Fermi-operators $\hat{a}^+$, $\hat{a}$.

Let us expand the operator $\hat{U}$ in Fermi-amplitudes and confine ourselves to the linear terms (because the perturbation is assumed to be sufficiently small)

$$\hat{U} \cong \mu_0(t) + \sum_n \alpha_n(t)\hat{a}_n + \sum_n \beta_n(t)\hat{a}_n^+ . \tag{16-16}$$

The requirement of hermiticity and unitarity of the operator $\hat{U}$ gives some additional conditions for the coefficients μ_0, α_n and β_n. Taking into account these conditions we can rewrite (16-16) in the following way

$$\hat{U} = \frac{1}{\lambda(t)}\left[1 + \sum_n \left(\alpha_n(t)\hat{a}_n^+ - \alpha_n^*(t)\hat{a}_n\right)\right], \tag{16-17}$$

where $|\lambda(t)|^2 = 1 + \sum_n |\alpha_n(t)|^2$.

Then the function of the state of the system at time t will be determined by the formula

$$\Psi(t) = \frac{1}{\lambda(t)}\left\{1 + \sum_n \left[\alpha_n(t)\hat{a}_n^{+} - \alpha_n^{*}(t)\hat{a}_n\right]\right\}|0\rangle . \tag{16-18}$$

The construction of the operator $\hat{U}$ and the function $\Psi(t)$ will be entirely completed if we determine the coefficients $\alpha_n(t)$. The physical sense of the coefficients can be easily established by formula (16-18). Indeed, it is not difficult to find that the quantity $\dfrac{\alpha_n(t)}{|\lambda(t)|^2}$ is interpreted as the amplitude of probability for the system to be in the state $\hat{a}_n^{+}|0\rangle$. Equations determining the coefficients $a_n(t)$ can be found from the quantum equations of motion for operators $\hat{a}_n$. To illustrate the last statement let us consider a simple example, when the dynamics of excited states of a system is described by quantum equations of the following type

$$i\frac{d\hat{a}_n}{dt} = \Delta\hat{a}_n + \mathcal{M}(\hat{a}_{n+1} + \hat{a}_{n-1}), \quad n = 1,2,\ldots N. \tag{16-19}$$

Let us multiplay eq. (16-19) by $\Psi(0)$ (from the right side) and by $\Psi(0)$ (from the left side), and integrate the result over the variable q. As a result we obtain

$$i\frac{d}{dt}\langle 0|\hat{a}_n(t)|0\rangle = \Delta\langle 0|\hat{a}_n(t)|0\rangle + \mathcal{M}\{\langle 0|\hat{a}_{n+1}(t)|0\rangle + \langle 0|\hat{a}_{n-1}(t)|0\rangle\}, \tag{16-20}$$

where $\langle 0|\hat{a}_n(t)|0\rangle \equiv \int \Psi^{*}(0)\hat{a}_n(t)\Psi(0)\,dq$.

Using the relation

$$\langle 0|\hat{a}_n(t)|0\rangle = \langle 0|\hat{U}^{+}(t)\hat{a}_n(0)\hat{U}(t)|0\rangle = \frac{\alpha_n(t)}{|\lambda(t)|^2}, \tag{16-21}$$

we can rewrite (16-20) in the following form

$$i\dot{\alpha}_n = \Delta \alpha_n + \mathcal{M}(\alpha_{n+1} - \alpha_{n-1}). \quad (16\text{-}22)$$

Thus we succeeded in obtaining the equation for the classical functions $\alpha(t)$. The solutions of the equation uniquely determine the function of the state of the model system $\Psi(t)$ (see (16-18)).

§17. Mathematical description of the continuous systems.

Till now we considered model systems belonging to the class of the so-called discrete systems, which have finite or numerable number of the degrees of freedom. In this and the next Parts, continuous systems where every point takes part in the motion will be considered. Sometimes the continuous systems are named also distributed systems. In this paragraph we shall describe the way in which the classical methods of description of the systems of many particles (the method of Lagrange and that of Hamilton) are modified in the case of continuous systems.

The simplest and generally accepted way consists in the following. At the beginning the continuous system is approximated by a discrete one, and the corresponding lagrangian (or hamiltonian) and equations of motion are constructed. Then the transformation from the discrete to the continuous system is carried out. First works using this way goes back to Rayleigh and their appearance is due to studying the problems of vibrations of strings, membranes and beams.

To illustrate the method let us consider as an example the vibrations of an infinitely long elastic rod. This continuous model can be approximated by a discrete system, namely by the chain consisting of N material particles (we shall name them the "molecules") of mass m, being a distance a apart and being connected with one another by springs with rigidity k (Fig.23). Let us assume also that these "molecules" can move only along the straight line which coincide with the chain. Equations describing oscillations of the "molecules" near their equilibrium positions have then the following simple form

$$m\ddot{q}_j = \kappa(q_{j+1} - q_j) - \kappa(q_j - q_{j-1}), \quad j = 1, 2, \ldots N, \tag{17-1}$$

where q_j is the displacement of the j-th "molecule" from its equilibrium position.

Let us rewrite these equations in the equivalent form

$$\frac{m}{a}\ddot{q}_j = \kappa a \frac{q_{j+1} - q_j}{a} - \kappa a \frac{q_j - q_{j-1}}{a}. \tag{17-2}$$

This form of the equations is more convenient in order to pass to the continuum limit, when $a \to 0$. If $a \to 0$, the terms of equation (17-2) tend to the following limits

$$q_j(t) \to q(x,t), \quad ja \to x, \quad \frac{m}{a} \to \rho, \quad \kappa a \frac{q_{j+1} - q_j}{a} \to Y\frac{\partial q}{\partial x},$$

$$\kappa a\left(\frac{q_{j+1} - q_j}{a} - \frac{q_j - q_{j-1}}{a}\right)/a \to Y\frac{\partial^2 q}{\partial x^2}, \tag{17-3}$$

where ρ is a linear density and Y is Young's modulus. Taking into account (17-3) we obtain the equations describing vibrations of a continuous elastic rod

$$\rho\frac{\partial^2 q}{dt^2} - Y\frac{\partial^2 q}{\partial x^2} = 0. \tag{17-4}$$

To find the function of Lagrange corresponding to equations (17-4) let us consider again the discrete analog which is the chain consisting of N "molecules". It is not difficult to show that the lagrangian corresponding to equations (17-1) has the following form

$$\mathcal{L} = \frac{1}{2} \sum_j \{ m\dot{q}_j^2 - \kappa (q_{j+1} - q_j)^2 \}. \tag{17-5}$$

Rewriting equation (17-5) in the equivalent form

$$\mathcal{L} = \frac{1}{2} \sum_j a \left\{ \frac{m}{a} \dot{q}_j^2 - \kappa a \left(\frac{q_{j+1} - q_j}{a} \right)^2 \right\}, \tag{17-6}$$

and proceeding in (17-6) to the continuous limit ($a \to 0$), we obtain the lagrangian of vibrating elastic rod

$$\mathcal{L} = \frac{1}{2} \int \left\{ \rho \left(\frac{\partial q}{\partial t} \right)^2 - Y \left(\frac{\partial q}{\partial x} \right)^2 \right\} dx. \tag{17-7}$$

The hamiltonian of the system

$$\mathcal{H} = \frac{1}{2} \int \left\{ \frac{p^2}{2\rho} + Y \left(\frac{\partial q}{\partial x} \right)^2 \right\} dx. \tag{17-8}$$

is obtained in an analogous way.

Analyzing equation (17-7) it is easy to note one peculiarity which consists in the presence of the derivative $\partial q/\partial x$ under integral expression. The latter is known as density of lagrangian ℓ and is determined by the equation

$$\mathcal{L} = \int l dx, \tag{17-9}$$

This peculiarity is typical of other continuous systems too. Summarizing the results, we can say that the density of the lagrangian of one-dimensional continuous system is a function not only of the variables q, t and $\partial q/\partial t$ but also of the derivative $\partial q/\partial x$

$$l = l\,(q, q_x, q_t, t). \tag{17-10}$$

This peculiarity leads to modification of equations of Lagrange (4-2) and they take the form

$$\frac{d}{dt}\frac{\partial l}{\partial q_t} + \frac{d}{dx}\left[\frac{\partial l}{\partial q_x}\right] - \frac{\partial l}{\partial q} = 0. \tag{17-11}$$

As in the case of discrete systems these equations are deduced from the general variational principle

$$\delta I = \delta \int_{t_1}^{t_2} \mathcal{L}\, dt = 0, \tag{17-12}$$

which is valid for both the discrete and the continuous systems.

In conclusion a few words must be said about modification of the dynamical equations in the case of the formalism of Hamilton. In this case the function of Hamilton is determined by the formula

$$\mathcal{H} = \int dx\,(p q_t - l), \tag{17-13}$$

where $p = \frac{\partial l}{\partial q_t}$ is the density of impulse. It is not difficult to make sure that the dynamical equations derived from the general variational principle have the following form

$$\frac{dp}{dt} = -\frac{\delta \mathcal{H}}{\delta q} = -\frac{\partial h}{\partial q} + \frac{d}{dx}\left[\frac{\partial h}{\partial(\partial q/\partial x)}\right], \qquad (17\text{-}14)$$

$$\frac{dq}{\partial t} = \frac{\delta \mathcal{H}}{\delta p} = \frac{\partial h}{\partial p},$$

where h is the density of the hamiltonian determined by the equation

$$\mathcal{H} = \int h\, dx\,. \qquad (17\text{-}15)$$

§18. Nonlinear Schrödinger equation.

In this and in the next two paragraphs a brief description of several nonlinear equations is given. The choice of the equations is determined by their applicability to studying the dynamics of biopolymers. We begin with the nonlinear Schrödinger equation

$$\varphi_{xx} + i\varphi_t + \kappa|\varphi|^2\varphi = 0, \tag{18-1}$$

where φ is a complex function of the variables x and t; k is a constant which is determined by parameters of the model system. Equation (18-1) is equivalent to the system of two equations

$$\begin{cases} \varphi_{xx} + i\varphi_t + \kappa|\varphi|^2\varphi = 0, \\ \varphi^*_{xx} - i\varphi^*_t + \kappa|\varphi|^2\varphi^* = 0. \end{cases} \tag{18-2}$$

These equations can be obtained from the lagrangian

$$\mathcal{L} = \int_{-\infty}^{+\infty} [(i/2)(\varphi^*\varphi_t - \varphi\varphi^*_t) - |\varphi_x|^2 + (\kappa/2)|\varphi|^4]\, dx \tag{18-3}$$

according to the rules

$$\begin{aligned} &\frac{d}{dt}\frac{\partial \ell}{\partial \varphi_t} + \frac{d}{dx}\left[\frac{\partial \ell}{\partial \varphi_x}\right] - \frac{\partial \ell}{\partial \varphi} = 0, \\ &\frac{d}{dt}\frac{\partial \ell}{\partial \varphi^*_t} + \frac{d}{dx}\left[\frac{\partial \ell}{\partial \varphi^*_x}\right] - \frac{\partial \ell}{\partial \varphi^*} = 0, \end{aligned} \tag{18-4}$$

which are the natural generalization of formula (17-11).

Determining the densities of impulses by the formulas

$$p = \frac{\partial \ell}{\partial \varphi_t} = (i/2)\varphi^*, \qquad p^* = \frac{\partial \ell}{\partial \varphi_t^*} = -(i/2)\varphi, \tag{18-5}$$

we can construct the hamiltonian of the system

$$\mathcal{H} = \int_{-\infty}^{+\infty} (p\varphi_t + p^*\varphi_t^* - \ell)\,dx = \int_{-\infty}^{+\infty} [|\varphi_x|^2 - (\kappa/2)|\varphi|^4]\,dx. \tag{18-6}$$

The nonlinear Schrödinger equation is widely used in physics to describe various phenomena: stationary two-dimensional self-focusing of plane waves, one-dimensional self-modulation of monochromatic waves, propagation of heat pulse in solids, propagation of Langmuir waves in plasma and other [43].

In mathematics, equation (18-1) is known as the S_3 equation. The letter S_3 symbolizes the first letter of the name of Schrödinger and the figure 3 indicates the cubic character of the nonlinearity. The S_3 equation belongs to a class of completely integrable equations which have an infinite number of so-called polynomial conservation laws. Usually these laws are written in the following form

$$\frac{\partial \rho_n}{\partial t} = -\frac{\partial j_n}{\partial x}, \quad n = 1, 2, \ldots, \tag{18-7}$$

where ρ_n is the linear density of some quantity being conserved and j_n is the flux of this quantity, which flows through a section (in the given one-dimensional case through a point) in unit time. Both quantities are polynomials in the variables φ, φ^*, φ_x and φ_x^*.

Integrating (18-7) over x and taking into account that $j_n \to 0$ when $x \to \pm\infty$, we can find the integral of motion I_n

$$I_n = \int_{-\infty}^{+\infty} \rho_n(x,t)\,dx. \tag{18-8}$$

The first three integrals of motion for the S_3 equation take the form

$$I_1 = \int_{-\infty}^{+\infty} dx\,|\varphi|^2,$$

$$I_2 = \int_{-\infty}^{+\infty} dx\,(\varphi^*\varphi_x - \varphi_x^*\varphi), \tag{18-9}$$

$$I_3 = \int_{-\infty}^{+\infty} dx\,[|\varphi_x|^2 - (\kappa/2)|\varphi|^4],$$

. . . .

They have a simple physical sense. Up to a multiplier they coincide with the number of particles, the impulse and the energy of the system, respectively.

Let us describe now one of the methods of finding the soliton solution of equation (18-1). For this purpose let us assume that the unknown complex function has the form

$$\varphi = \Phi(x,t)\exp[i\,\Theta(x,t)], \tag{18-10}$$

where Φ and Θ are real functions named the envelope and the carrier. The equations determining these functions can be found by inserting (18-10) into (18-1)

$$\Phi_{xx} - \Phi\Theta_x^2 - \Phi\Theta_t + \kappa\Phi^3 = 0,$$
$$\Phi\Theta_{xx} + 2\Phi_x\Theta_x + \Phi_t = 0. \tag{18-11}$$

If we assume that the solutions have the form of travelling waves

$$\Phi = \Phi(x - vt) = \bar{\Phi}(x), \qquad \Theta = \Theta(x - vt) = \bar{\Theta}(x), \tag{18-12}$$

where v and u are the velocities of envelope and carrier, respectively, equations (18-11) can be rewritten in the form of ordinary differential equations

$$\frac{d^2\bar{\Phi}}{dx^2} - \bar{\Phi}\left(\frac{d\bar{\Theta}}{dx}\right)^2 + u\,\bar{\Phi}\,\frac{d\bar{\Theta}}{dx} + \kappa\bar{\Phi}^3 = 0, \tag{18-13}$$

$$\bar{\Phi}\,\frac{d^2\bar{\Theta}}{dx^2} + 2\,\frac{d\bar{\Theta}}{dx}\cdot\frac{d\bar{\Phi}}{dx} - v\,\frac{d\bar{\Phi}}{dx} = 0. \tag{18-14}$$

Multiplying equation (18-14) by $\bar{\Phi}$ we can transform it to the form

$$\frac{1}{2}\frac{d}{dx}\left\{\bar{\Phi}^2\left(2\,\frac{d\bar{\Theta}}{dx} - v\right)\right\} = 0, \tag{18-15}$$

whence we find

$$\bar{\Phi}^2\left(2\,\frac{d\bar{\Theta}}{dx} - v\right) = Const. \tag{18-16}$$

Assuming that the boundary condition $\bar{\Phi}(\pm\infty) = 0$ is fulfilled we obtain the solution for the first derivative of the carrier

$$\frac{d\bar{\theta}}{dx} = \frac{v}{2}. \qquad (18\text{-}17)$$

Now let us return to equation (18-13), insert the solution (18-17) into the equation

$$\frac{d^2\bar{\Phi}}{dx^2} - \bar{\Phi}\left(\frac{v^2}{4} - \frac{uv}{2}\right) + \kappa\bar{\Phi}^3 = 0, \qquad (18\text{-}18)$$

and rewrite the obtained expression in the old terms ($\bar{\Phi} \to \Phi$)

$$\frac{d^2\Phi}{d\xi^2} - \Phi\left(\frac{v^2}{4} - \frac{uv}{2}\right) + \kappa\Phi^3 = 0, \qquad (18\text{-}19)$$

where $\xi = x - vt$.

To solve equation (18-19) let us multiply it by $\frac{d\Phi}{d\xi}$ and transform to the following form

$$\frac{d}{d\xi}\left[\left(\frac{d\Phi}{d\xi}\right)^2 - \Phi^2\left(\frac{v^2}{4} - \frac{uv}{2}\right) + \frac{\kappa}{2}\Phi^4\right] = 0, \qquad (18\text{-}20)$$

whence we find

$$\left(\frac{d\Phi}{d\xi}\right)^2 - \Phi^2\left(\frac{v^2}{4} - \frac{uv}{2}\right) + \frac{\kappa}{2}\Phi^4 = \overline{Const}. \qquad (18\text{-}21)$$

Assuming that the boundary condition $\left.\frac{d\Phi}{d\xi}\right|_{\xi=\pm\infty} = 0$ is fulfilled we find that $\overline{Const} = 0$. Then equation (18-21) can be transformed to

$$\frac{d\Phi}{\sqrt{P(\Phi)}} = d\xi ,$$

(18-22)

where the polynomial $P(\Phi)$ is determined by

(18-23)

$$P(\Phi) = \Phi^2\left(\frac{v^2}{4} - \frac{uv}{2}\right) - \frac{k}{2}\Phi^4 .$$

The function $P(\Phi)$ is shown in Fig. 24. Integrating (18-22) between 0 and $\Phi_0 = \sqrt{(v^2-2uv)/2k}$, we find the solution for envelope

$$\Phi = \Phi_0 / ch\,[\Phi_0 \sqrt{k/2}\,(x - vt)] . \qquad (18\text{-}24)$$

and the complete solution of the S_3 equation has then the form

$$\varphi = \Phi e^{i\theta} = \Phi_0 \exp[i(v/2)(x - vt)]/ch\,[\Phi_0 \sqrt{k/2}\,(x - vt)] . \qquad (18\text{-}25)$$

A schematic picture of the function φ is shown in Fig. 25. It looks like a wave packet with the envelope having the form of hyperbolic secant and moving with the velocity v, and with the carrier having the form of monochromatic wave moving with the velocity u.

Soliton solution of the S_3 equation can be obtained by another method which is known as the inverse scattering method. Zakharov and Shabat [55] used this method to find one-soliton solution

$$\varphi = \sqrt{2k}\,\eta \exp[-4i(\xi^2 - \eta^2)t - 2i\xi x + i\delta]/ch\,[2\eta(x - x_0) + 8\eta\xi t] ,$$

(18-26)

where ξ, η, δ and x_0 are free parameters, -4ξ being equal to v. The slight difference in the forms of equation (18-26) and (18-25) is explained by the symmetry of the scattering problem. The reader can find detailed informations about the inverse scattering method, and the transformation of Backlund permitting us to obtain a set of other (many-soliton) solutions, in the references mentioned at the beginning of this Part. However, in applications only one-soliton solutions of type (18-25) are usually used. They can be interpreted as amplitudes of probability of the existence of localized (in the region of the length $d \cong \pi\sqrt{2}/\varphi_0\sqrt{k}$) excitations moving with the velocity v.

§19. Sine-Gordon equation.

The sine-Gordon equation, which has the form

$$\varphi_{xx} - \varphi_{tt} = \sin\varphi \tag{19-1}$$

can be obtained from the lagrangian.

$$L = \int_{-\infty}^{+\infty} \left[\frac{1}{2} (\varphi_t^2 - \varphi_x^2) + \cos\varphi \right] dx . \tag{19-2}$$

Here (in contrast to the previous paragraph) φ is a real function of the two variables x and t.

If we define the density of impulse p by the equation

$$p = \frac{\partial l}{\partial \varphi_t} = \varphi_t , \tag{19-3}$$

the hamiltonian constructed according to rule (17-13), will have the following form

$$H = \int_{-\infty}^{+\infty} (p\varphi_t - l)dx = \int_{-\infty}^{+\infty} \left[\frac{1}{2} (\varphi_t^2 + \varphi_x^2) - \cos\varphi \right] dx \tag{19-4}$$

or it can be rewritten (up to a constant) in the more convenient form

$$H = \int_{-\infty}^{+\infty} \left[\frac{1}{2} (\varphi_t^2 + \varphi_x^2) + (1 - \cos\varphi) \right] dx. \tag{19-5}$$

The sine-Gordon equation finds an application in various physical problems. Propagation of dislocations in crystals, motion of Bloch walls in magnetic crystals, propagation of magnetic flux on a Josephson line and a unitary theory for elementary particles are among them [43]. Note that the possibility to use the sine-Gordon equation in the theory of elementary particles is associated with the property of invariance under Lorentz transformation

$$x \to x' = \mu\,(x - vt), \qquad t \to t' = \mu\,(t - vt), \qquad (19\text{-}6)$$

where $\mu = (1 - v^2)^{-1/2}$.

In mathematics, equation (19-1) is often named the sG equation. The small letter s symbolizes the first letter of the word sine and the capital letter G the first letter of the name of Gordon. As in the case of the S_3 equation, the sG equation belongs to a class of completely integrable equations which have an infinite system of conservation laws of type (18-8). To illustrate this statement we present here several conserved densities ρ_n

$$\rho_1 = \frac{1}{2}\varphi_x^2,$$

$$\rho_2 = \frac{1}{2}\varphi_{xx}^2 - \frac{1}{8}\varphi_x^4, \qquad (19\text{-}7)$$

$$\rho_3 = \frac{1}{2}\varphi_{xxx}^2 - \frac{5}{4}\varphi_{xx}^2\varphi_x^2 + \frac{1}{16}\varphi_x^6,$$

. . . .

It can be easily shown that the density ρ_1 is equivalent to the density of the hamiltonian, h, and hence the first integral

$$I_1 = \int_{-\infty}^{+\infty} \rho_1\, dx \qquad (19\text{-}8)$$

represent the law of energy conservation.

Now let us find soliton solution of the sG equation. The scheme for finding the solution is very similar to that considered in the previous paragraph. We shall begin with the assumption that the unknown solution has the form of travelling wave

$$\varphi = \varphi(x - vt), \tag{19-9}$$

where v is the velocity of the wave. Inserting (19-9) into (19-1) we obtain an ordinary differential equation

$$\frac{d^2\varphi}{d\xi^2}(1 - v^2) = \sin\varphi, \tag{19-10}$$

where $\xi = x - vt$. To solve equation (19-10) let us multiply it by $d\varphi/d\xi$ and transform the result to the form

$$\frac{d}{d\xi}\left[\frac{1}{2}\left(\frac{d\varphi}{d\xi}\right)^2(1 - v^2) + \cos\varphi\right] = 0. \tag{19-11}$$

Integrating (19-11) and taking into account the boundary conditions $\left.\frac{d\varphi}{d\xi}\right|_{\xi = \pm\infty} = 0$, we obtain

$$\frac{d\varphi}{\sqrt{P(\varphi)}} = \frac{d\xi}{\sqrt{1 - v^2}}, \tag{19-12}$$

where $P(\varphi) = 2(1 - \cos\varphi)$. A schematic picture of the function $P(\varphi)$ is shown in Fig. 26. Integrating equation (19-12) we find at last the unknown solution

$$\varphi = 4\,\mathrm{arctg}\,\{\exp \pm [\mu(x - vt - x_0)]\}. \tag{19-13}$$

This solution taken with the sign plus is named kink. It satisfies the boundary conditions: $\lim_{\xi \to -\infty} \varphi = 0$ and $\lim_{\xi \to +\infty} \varphi = 2\pi$. The solution (19-13) taken with the sign minus is named antikink and it has the asymptotes: $\lim_{\xi \to -\infty} \varphi = 2\pi$ and $\lim_{\xi \to +\infty} \varphi = 0$. Schematic pictures of kink and antikink are presented in Fig. 27. The peculiarity of the kink- (or antikink-) solutions is that they are not localized functions but their derivatives φ_ξ are localized ones. Besides that kink- and antikink-solutions are an example of topological solitons. The topological charge Q

$$Q = \frac{1}{2\pi} \int_{-\infty}^{+\infty} \varphi_x \, dx . \tag{19-14}$$

is a quantity which characterizes the topological properties of the solutions. It is equal to +1 and −1 for kinks and antikinks, respectively.

Solutions (19-13) in the form of kink and antikink are the one-soliton solutions of the sG equation. As in the previous paragraph we can confine ourselves to consideration of this type of the solutions because only these will be used in applications.

A remarkable feature of the sG equation is the possibility to construct a simple mechanical model which can visually demonstrate the properties of the solutions. For the first time this model was described and recommended by Scott [56] as illustrative material for the teaching of students. A schematic picture of the mechanical model is shown in Fig. 28. It represents a long chain of pendula hanging up at a distance a apart and coupling with one another by springs with the rigidity k. To show that this mechanical system is really described by the sG equation it is enough to construct the equations describing torsional oscillations of pendula

$$I\ddot{\varphi}_n = K(\varphi_{n+1}-\varphi_n) - K(\varphi_n-\varphi_{n-1}) - T\sin\varphi_n . \qquad (19\text{-}15)$$

Here I is the moment of inertia of a pendulum; φ_n is the angular displacement of the n-th pendulum from its equilibrium position; $T\sin\varphi_n$ is the force moment acting on the n-th pendulum due to the Earth gravitational field. If we denote by the letters l, m and g the length, mass and gravitational constant, respectively, then we can write that $I=ml^2$ and $T=mg$.

Passing to the continuum limit ($a\to 0$) in (19-15) we obtain

$$J\varphi_{tt} = Y\varphi_{xx} - t_0 \sin\varphi, \qquad (19\text{-}16)$$

where $J=I/a$, $t_0=T/a$, and Y is the Young's modulus.

At last, by the transformation of variables

$$x \to x' = \sqrt{K/T}\, x ,$$
$$t \to t' = \sqrt{J/T}\, t \qquad (19\text{-}17)$$

equation (19-16) can be reduced to canonical form (19-1). This result proves the equivalency of the mechanical model to the sG equation. A qualitative picture that can be observed in the mechanical model is shown in Fig. 29. It is assumed there that the initial amplitude of the angular displacement of one of the pendula is rather large and the boundary conditions are $\left.\frac{\partial\varphi}{\partial x}\right|_{x=0} = \left.\frac{\partial\varphi}{\partial x}\right|_{x=L} = 0$.

§20. Other nonlinear equations.

In this paragraph we present a brief description of another two nonlinear equations which the reader will meet in applications. They are the equation of φ^4

$$\varphi_{xx} - \varphi_{tt} = \pm (\varphi - \varphi^3), \tag{20-1}$$

which received this name due to the hamiltonian

$$\mathcal{H} = \int_{-\infty}^{+\infty} [(\varphi_t^2 + \varphi_x^2)/2 \pm (\varphi^2 - \varphi^4/2)/2]\, dx, \tag{20-2}$$

and the double sine-Gordon equation

$$\varphi_{xx} - \varphi_{tt} = \sin\varphi + 2\lambda \sin 2\varphi, \tag{20-3}$$

with hamiltonian of the following form

$$\mathcal{H} = \int_{-\infty}^{+\infty} [(\varphi_t^2 + \varphi_x^2)/2 + (2 - \cos\varphi - \lambda \cos 2\varphi)]dx. \tag{20-4}$$

The double sine-Gordon equation is named often in mathematics the dsG equation. The additional small letter d in the name symbolizes the first letter in the word "double".

In contrast to the S_3 and the sG equations, the φ^4 and the dsG ones do not belong to the class of the completely integrable equations. An important feature of these two equations is that they belong to the class of equations of the type

$$\varphi_{tt} - \varphi_{xx} + \frac{\partial V}{\partial \varphi} = 0, \tag{20-5}$$

which are known as the Klein-Gordon equations. The hamiltonian corresponding to the equations has the form

$$\mathcal{H} = \int_{-\infty}^{+\infty} [(\varphi_t^2 + \varphi_x^2)/2 + V(\varphi)]\, dx , \tag{20-6}$$

where $V(\varphi)$ is a dimensionless potential. Notice that the sG equation described in the previous paragraph also belongs to this class. So, the potential $V(\varphi)$ for the sG equation is equal to

$$V(\varphi) = 1 - \cos\varphi, \tag{20-7}$$

for the φ^4 equation V( ) takes the form

$$V(\varphi) = \pm(\varphi^2 - \varphi^4/2)/2, \tag{20-8}$$

and for the dsG equation the potential is

$$V(\varphi) = 2 - \cos\varphi - \lambda \cos 2\varphi . \tag{20-9}$$

A schematic picture of different potentials $V(\varphi)$ is shown in Fig. 30. Using the scheme of calculation described in the previous paragraphs one can find soliton solutions of equations (20-1) and (20-3). So, the equation φ^4 taken with the sign plus has the solitary wave solution

$$\varphi = \pm \operatorname{sech} [\mu(x - vt)/\sqrt{2}]. \tag{20-10}$$

The same equation but taken with the minus sign has the solutions in the forms of kink and antikink

$$\varphi = \pm\, th\left[\mu(x - vt)\right]. \tag{20-11}$$

Just this case is used in applications. A schematic picture of the function φ determined by formula (20-11), is shown in Fig. 31.

Equation (20-3) also has the solutions of kink and antikink types

$$\varphi = 2\, arctg\left\{\pm\left(\sqrt{1+4\lambda}\right)^{-1} sh^{-1}\left[\mu\sqrt{1+4\lambda}\,(x - vt + x_0)\right]\right\}. \tag{20-12}$$

It can be shown that these solutions are valid under conditions usually occurring in applications. The general form of the function (20-2) is shown in Fig. 32. We shall not give here the other solutions of the dsG equation. The reader interested in details can find them in the review of Condat and coauthors [57].

§21. Perturbation theory for solitons.

The equations S_3, sG, dsG and φ^4 considered in the previous paragraphs are suitable for the description of different nonlinear processes but only in the first, very rough approximation. Any attempt to improve the description leads to the necessity to modify the dynamical equations by introducing additional terms. They can be the terms imitating the effects of dissipation, external fields, impurities and so on. In the case, when the additional terms are small corrections to the basic (unperturbed) equation, the solutions of modified equations can be found by the methods of the theory of perturbation. In this paragraph we shall illustrate application of the method of the perturbation theory to the problem of finding the solutions of modified nonlinear equations.

For definiteness let us assume that unperturbed equation is the sG equation

$$\varphi_{tt} - \varphi_{xx} + \sin\varphi = 0. \tag{21-1}$$

Assume also that V is a small correction (or perturbation); then the complete (or perturbed) problem takes the following form

$$\varphi_{tt} - \varphi_{xx} + \sin\varphi - V = 0 \tag{21-2}$$

Note that in the general case V can be a function of variables x, t, φ, φ_x, φ_t and so on.

The solution of problem (21-2) can be found in the form

$$\varphi = \varphi^{(0)} + \varphi^{(1)}, \tag{21-3}$$

where $\varphi^{(0)}$ is the one-soliton solution of the unperturbed equation (21-1) which is determined by formula (19-3); $\varphi^{(1)}$ is a small correction. The problem consists in finding the function $\varphi^{(1)}(x,t)$.

The equation for the function $\varphi^{(1)}$ can be found by inserting (21-3) into (21-2)

$$\varphi^{(1)}_{tt} - \varphi^{(1)}_{xx} + \cos\varphi^{(0)} \cdot \varphi^{(1)} - V = 0. \tag{21-4}$$

It is convenient to transform (21-4) to the new variables

$$x \to x' = \mu(x - vt), \quad t \to t' = \mu(t - vx), \tag{21-5}$$

where $\mu = (1-v^2)^{-1/2}$. For this purpose, let us write the rules by which the functions of equation (21-4) and their derivatives are transformed

$$\varphi^{(0)} \to \varphi'^{(0)}, \quad \varphi^{(1)} \to \varphi'^{(1)}, \quad V \to V',$$

$$\varphi^{(1)}_{tt} \to \mu^2 [\varphi'^{(1)}_{t't'} - 2v\varphi'^{(1)}_{t'x'} + v^2\varphi'^{(1)}_{x'x'}], \tag{21-6}$$

$$\varphi^{(1)}_{xx} \to \mu^2 [\varphi'^{(1)}_{x'x'} - 2v\varphi'^{(1)}_{x't'} + v^2\varphi'^{(1)}_{t't'}].$$

Then insert (21-6) into (21-4). As a result we obtain

$$\varphi'^{(1)}_{t't'} - \varphi'^{(1)}_{x'x'} + \varphi'^{(1)}\cos\varphi'^{(0)} = V'. \tag{21-7}$$

Note that the transformation used is equivalent to the transition to a new system of coordinates moving with a constant velocity v.

To find the solution of inhomogeneous equation (21-7) we can use the following method. First let us consider the homogeneous equation

$$\varphi'^{(1)}_{t't'} - \varphi'^{(1)}_{x'x'} + \varphi'^{(1)} \cos \varphi'^{(0)} = 0. \qquad (21-8)$$

Assume that its solution can be written in the form

$$\varphi'^{(0)} = e^{iWt'} f(x'), \qquad (21-9)$$

where W=Const. Inserting (21-9) into (21-8), we find the equation for the function f

$$-f_{x'x'} + f \cos \varphi'^{(0)} = W^2 f. \qquad (21-10)$$

It becomes obvious that (21-10) is the problem of finding eigenvalues and eigenfunctions of the operator

$$\hat{\mathcal{D}} = -\frac{d^2}{dx'^2} + \cos \varphi'^{(0)}. \qquad (21-11)$$

The solutions of the problem were investigated in [58, 59]. They consists of:

1) one bound state which is characterized by the eigenfunction

$$f_b(x') = 2\,\mathrm{sch}\,x' \qquad (21-12)$$

and by the eigenvalue

$$W_b^2 = 0, \qquad (21-13)$$

2) the continuum spectrum which is characterized by the ensemble of the eigenfunctions

$$f_k(x') = \sqrt{2\pi}\, e^{ikx'} (k + i\,th\,x') \tag{21-14}$$

and the eigenvalues

$$W_k^2 = 1 + k^2. \tag{21-15}$$

Because the functions f_b and f_k are the eigenfunctions of the self-adjoint operator $\hat{D}$, then according to the theorems of linear algebra these functions form a complete orthonormal basis. The condition of orthogonality is provided by the equations

$$\int_{-\infty}^{+\infty} f_b(x)\, f_b(x)\, dx = 8\,,$$

$$\int_{-\infty}^{+\infty} f_k^*(x)\, f_{k'}(x)\, dx = \delta(k-k') \tag{21-16}$$

and the condition of completeness is provided by the equation

$$\int_{-\infty}^{+\infty} f_k^*(x') f_k(x)\, dk + \frac{1}{8} f_b(x') f_b(x) = \delta(x'-x). \tag{21-17}$$

Let us return to the initial inhomogeneous equation (21-7) and expand its solution ($\varphi^{(1)}(x',t')$) in the base functions f_b, f_k

$$\varphi^{(1)}(x',t') = \frac{1}{8}\psi_b(t') f_b(x') + \int_{-\infty}^{+\infty} \psi_k(t') f_k(x')\, dk. \tag{21-18}$$

Thus to solve the problem it is now enough to find only the coefficients of the expansion: $\Psi_b(t')$ and $\Psi_k(t')$.

To find the equation determining the coefficient $\Psi_b(t')$ let us expand all terms of equation (21-7) in the base functions $f_b(x')$, $f_k(x')$, multiply the result by $f_b(x')$ and integrate over the variable x' between $-\infty$ and $+\infty$. As a result we obtain the equation

$$\Psi_{b_{t't'}} = q_b(t'). \tag{21-19}$$

Using the same method we can obtain the equation determining the coefficient $\Psi_k(t')$

$$\Psi_{k_{t't'}} + W_k^2 \Psi_k = q_k(t'). \tag{21-20}$$

The functions $q_b(t')$, $q_k(t')$ appearing on the right of equations (21-19) and (21-20) are the coefficients of the expansion of the perturbation V

$$V = \frac{1}{8} q_b(t') f_b(x') + \int_{-\infty}^{+\infty} q_k(t') f_k(x') dx'. \tag{21-21}$$

They can be found from the simple formulas

$$q_b(t') = \int_{-\infty}^{+\infty} V f_b(x') dx', \tag{21-22}$$

$$q_k(t') = \int_{-\infty}^{+\infty} V f_k(x') dx'. \tag{21-23}$$

Equations (21-19), (21-20) supplemented by (21-22) and (21-23) entirely complete the solution of the problem.

The method described above is rather general and can be applied to any nonlinear model. The only change that must be expected when we apply the method to some other nonlinear problems is the change of the operator $\hat{D}$ and, hence, of the form of the eigenfunctions f_b, f_k and their eigenvalues W_b^2, W_k^2.

For illustration of the technique of perturbation theory let us consider a simple example when V is a small constant equal to $-V_0$. Let us find the coefficient $q_b(t')$ from formula (21-22)

$$q_b(t') = \int_{-\infty}^{+\infty} V f_b(x')\,dx' = -2\pi V_0 , \tag{21-24}$$

insert it into equation (21-19)

$$\varphi_{b_{t't'}} = -2\pi V_0 , \tag{21-25}$$

and integrate the result over the variable t'

$$\varphi_{b_{t'}} = -2\pi V_0 t' + a , \tag{21-26}$$

where a is the constant which is determined by the initial conditions. Assuming then that $\varphi_{b_{t'}}(t'=0)=0$ we finally find

$$\varphi_{b_{t'}}(t') = -2\pi V_0 t'. \tag{21-27}$$

To understand the physical sense of the obtained result let us return to the beginning of the paragraph. At the very beginning one could imagine that under the action of perturbation V the form and the velocity of the soliton of the sG equation must be changed. Let us assume that these changes have an additive character and consider separately the problem of changing the velocity of the soliton when the form conserves unchanged. That is let us consider the case when

$$\varphi'(x',t') \cong \varphi'^{(0)}(x' - \Delta v(t')\cdot t'), \tag{21-28}$$

where $\Delta v(t')$ is the change of the velocity of the soliton due to the influence of perturbation. Here we takes into account that at the initial moment the soliton is at rest (in the system of coordinates $\{x', t'\}$, which is moving with the velocity v).

Then let us expand (21-28) in Taylor series

$$\varphi'(x',t') = \varphi'^{(0)}(x') - f_b(x')\,[\Delta v(t')\cdot t'] + \cdots \quad . \tag{21-29}$$

Here we take into account the equality

$$f_b(x') = \frac{d\varphi'^{(0)}}{dx'}, \tag{21-30}$$

which can be easily checked by differentiation of equation (21-1) and comparison of the result obtained with (21-10).

From (21-27), (21-3) and (21-18) we find the unknown formula for $\Delta v(t')$

$$\Delta v(t') \cong -\frac{1}{8}\,\frac{\Psi_b(t') - \Psi_b(0)}{t' - 0} \cong -\frac{1}{8}\,\Psi_b(t') = \frac{\pi V_0 t'}{4}, \tag{21-31}$$

It follows from this that the movement of the soliton under the action of small constant perturbation is uniformly accelerated. This result points at the similarity in the behavior of the soliton and in that of usual classical particle. This similarity is typical not only of the soliton of the sG equation, but of the solitons of other nonlinear equations too. On these grounds the solitons are often called "particle-like" excitations.

§ 22. Soliton statistics.

In the previous paragraphs we considered several nonlinear model systems: S_3, φ^4, sG, sG with perturbation and dsG. In all these models it was suggested that the total number of nonlinear waves (solitons) is equal to 1 or that it is so small that the collisions and interactions between solitons can be neglected. In this paragraph we shall try to take off this limitation and assume that the total number of solitons N_S is a macroscopic quantity. This leads naturally to the problem of the modelling of statistical ensembles of solitons.

In the general case the solution of the problem is rather complex, and here we shall give a brief description of only one of the simple approximate methods of finding the solution. This method is a phenomenological one. It is based on the similarity between the properties of solitons and ordinary classical particles. Indeed, just as in the case of classical material particle, we can ascribe to the soliton some mass and velocity. For example, let us consider the model sG. Using formulas (19-5) and (19-13) we can calculate the soliton energy $\mathcal{E}_S$

$$\mathcal{E}_S = \int_{-\infty}^{+\infty} [(\varphi_t^2 + \varphi_x^2)/2 + (1 - \cos\varphi)]\, dx = 8\mu. \qquad (22\text{-}1)$$

In the "nonrelativistic" limit when the velocity of soliton v is small, equation (22-1) takes the form

$$\mathcal{E}_S \cong 8\left(1 + \frac{v^2}{2}\right). \qquad (22\text{-}2)$$

This result can be interpreted as a sum of the kinetic ($T=8v^2/2$) and the potential ($V=8$) energies. Thus we can ascribe mass $m=8$ and the velocity v to the nonrelativistic soliton of the equation sG. In the relativistic case the soliton of the sG equations can be characterized by the mass $m=m_0\gamma$, impulse $p=8\gamma v$ and the energy $\mathcal{E}_S=\mathcal{E}_0\gamma$. Here m_0 is the mass of rest ($m_0=8$), $\mathcal{E}_0$ is the energy of rest ($\mathcal{E}_0=8$).

Besides that, the reader has had already an opportunity to make sure that the behavior of a soliton moving under the action of an external constant field is very similar to the behavior of material particle. An analogous conclusion follows from the analysis of the stability of solitons and from the character of the interaction of solitons with one another (see, for example, [43]).

All these considerations suggest us to model an ensemble of solitons in the form of ordinary classical system consisting of N_S interacting material particles which have mass, impulse and energy of solitons. If the number N_S is not large the interactions between the particles can be neglected, and the model comes to a simpler variant, namely to an "ideal gas" of solitons.

To obtain different macroscopic characteristics of the excited nonlinear system it remains only to use the results of the classical statistical physics for the model of the ideal gas of N_S particles. Taking into account that N_S is not a fixed number let us write the large statistical sum in the form

$$\Xi(T,L,\mu)=\sum_{N_S=0}^{\infty} e^{\mu N_S/k_B T} z_{N_S}=\sum_{N_S=0}^{\infty}\frac{1}{N_S!}\left[e^{\mu/k_B T} z_0\right]^{N_S} \quad (22\text{-}3)$$

$$=\exp\left[e^{\mu/k_B T} z_0\right],$$

where

$$z_0 = \frac{1}{2\pi}\int_0^L dx \int dp\, e^{-\mathcal{E}_s/\kappa_B T} = \frac{L}{2\pi}\int dp\, e^{-\mathcal{E}_s/\kappa_B T}.$$

With the help of Ξ we can calculate various equilibrium characteristics of the model system, namely the thermodynamic potential G, the density of solitons n_S, the capacity C_L and others. For illustration let us give here a scheme of calculation of the density of solitons (kinks) of the sG model.

Firstly let us calculate the quantity z_0

$$z_0 = \frac{L}{2\pi}\int dp\, e^{-\mathcal{E}_s/\kappa_B T} = \frac{8L}{\pi}\mathcal{K}_1\left(\frac{8}{\kappa_B T}\right), \qquad (22\text{-}4)$$

where $\mathcal{K}_1(x)$ is the function of McDonald. Inserting (22-4) into (22-3), we find the large statistical sum Ξ

$$\Xi = \exp\left[e^{\mu/\kappa_B T}\cdot\frac{8L}{\pi}\cdot\mathcal{K}_1\left(\frac{8}{\kappa_B T}\right)\right]. \qquad (22\text{-}5)$$

Let us use then the formulas of classical statistical physics, according to which the density of particles is determined by the equation

$$n_s = -\frac{1}{L}\left.\frac{\partial G}{\partial \mu}\right|_{\mu=0}, \qquad (22\text{-}6)$$

where G is the thermodynamic potential. The latter is in turn determined by the equation

$$G = -\kappa_B T \ln \Xi. \qquad (22\text{-}7)$$

Inserting (22-5) and (22-7) into (22-6) we finally obtain

$$n_s = \frac{8}{\pi}\mathcal{K}_1\left(\frac{8}{\kappa_B T}\right). \qquad (22\text{-}8)$$

In the case of "low" temperature ($T \ll \frac{}{k_B}$) formula (22-8) comes to the simpler form

$$n_s \cong 2\sqrt{k_B T/\pi}\, \exp(-8/k_B T). \qquad (22\text{-}9)$$

PART 5. APPLICATIONS. INTERNAL DYNAMICS OF BIOPOLYMERS.

In this Part applications of the theory of nonlinear waves to studying the dynamical properties of biopolymers are considered. Many new results including those obtained by the author are presented. The solutions of the problems of energy migration in alpha-helical proteins [54], of scattering of neutrons and light by the nonlinear excitations in DNA [60], of dissipation and inhomogeneity in the models of nonlinear dynamics of biopolymers [61], of constructing more realistic model of the DNA dynamics [62] are among them.

§23. Transmission of nonlinear waves in molecular chains.

In this paragraph we consider again alpha-helical proteins with the structure described in details in §2. We shall deal with the problem of propagation of excitation waves along the alpha-helical proteins. The method of solving the problem contains two important steps. The first is the modeling of the dynamical equations, the second is the finding of their solutions and interpretation of them. In this and in the next paragraphs we shall obtain the equations describing the dynamics of the excited states of alpha-helical proteins, solve them and discuss a possible interpretation of the solutions.

Physical models of alpha-helical proteins have been discussed already in Part 3. At the beginning let us use the simplest of them, namely, the model of the chain of identical "molecules", and then generalize the obtained results and extend them to the more complex model of Suprun and Davydov.

A schematic picture of the chain of identical "molecules" is shown in Fig. 14, and the model hamiltonian $\hat{\mathcal{H}}$ is presented by formula (10-2) or by the equivalent one (10-4). However, when constructing the hamiltonian $\hat{\mathcal{H}}$ it was assumed that the average number of excitations $\langle \hat{N} \rangle$ was small. On these grounds the quartic term having the form

$$\sum_n Z \hat{N}_n \hat{N}_{n+1} , \tag{23-1}$$

and appearing in the initial model (10-1) was omitted. There is one more reason for the appearance of the quartic terms in the model hamiltonian. So, in §5 it has been shown that the quartic term

$$\sum_n h \hat{N}_n \hat{N}_{n+1} , \tag{23-2}$$

can appear as a result of interaction of "intramolecular" excitations with the lattice oscillations. To take into account both these effects, let us modify hamiltonian (10-4) by adding the quartic term

$$\hat{\mathcal{H}} = \sum_n \Delta \hat{a}_n^+ \hat{a}_n + \sum_n \mathcal{M} (\hat{a}_n^+ \hat{a}_{n+1} + \hat{a}_{n+1}^+ \hat{a}_n) + \\ + \sum_n \mathcal{M}' (\hat{a}_n^+ \hat{a}_{n+1}^+ - \hat{a}_n \hat{a}_{n+1}) + \sum_n Z \hat{N}_n \hat{N}_{n+1} . \tag{23-3}$$

Here the coefficients Δ, M, M′ and Z are determined as in §10.

To obtain the dynamical equations let us use the standard method of quantum statistical theory, which was used by us many times in Part 3. As a result we obtain the equation for the operators $\hat{a}_n$

$$i\frac{d\hat{a}_n}{dt} = \Delta\hat{a}_n + \mathcal{M}(\hat{a}_{n+1} + \hat{a}_{n-1}) + \mathcal{M}'(\hat{a}^+_{n+1} - \hat{a}^+_{n-1}) + \\ + \mathcal{Z}(\hat{N}_{n+1} + \hat{N}_{n-1})\hat{a}_n, \quad n = 1, 2, \ldots N. \tag{23-4}$$

In the quantum statistical theory the operator equations of type (23-4) are only intermediate results which are used to obtain the equilibrium characteristics of the system. In the dynamical theory these equations are of independent importance. As a matter of fact they are the dynamical equations sought for.

There are two reasons why the finding of analytical solutions of equations (23-4) seems a rather difficult problem. First, these equations are operator ones; second, they are discrete ones. Both these difficulties can be overcome if we use the $\hat{a} \to \Psi$ transformation and then proceed to the continuum limit.

So, let us begin by rewriting equations (23-4) in terms of classical function. Following the method described in §16, let us project equation (23-4) to the vacuum state $|0\rangle$

$$i\frac{d}{dt}\langle 0|\hat{a}_n(t)|0\rangle = \Delta\langle 0|\hat{a}_n(t)|0\rangle + \mathcal{M}\{\langle 0|\hat{a}_{n+1}(t)|0\rangle + \\ + \langle 0|\hat{a}_{n-1}(t)|0\rangle\} + \mathcal{M}'\{\langle 0|\hat{a}^+_{n+1}(t)|0\rangle - \langle 0|\hat{a}^+_{n-1}(t)|0\rangle\} + \\ + \mathcal{Z}\{\langle 0|\hat{N}_{n+1}(t)\hat{a}_n(t)|0\rangle + \langle 0|\hat{N}_{n-1}(t)\hat{a}_n(t)|0\rangle\}. \tag{23-5}$$

Let us apply an additional procedure of uncoupling to the tertiary terms

$$\langle 0|\hat{N}_m(t)\hat{a}_n(t)|0\rangle \cong \langle 0|\hat{a}^+_m(t)|0\rangle\langle 0|\hat{a}_m(t)|0\rangle\langle 0|\hat{a}_n(t)|0\rangle \tag{23-6}$$

and introduce new notations

$$\langle 0|\hat{a}_n(t)|0\rangle = \langle 0|\hat{U}^+(t)\hat{a}_n\hat{U}(t)|0\rangle = \frac{\alpha_n(t)}{|\lambda(t)|^2} \equiv \varphi_n(t),$$

$$\langle 0|\hat{a}_n^+(t)|0\rangle = \langle 0|\hat{U}^+(t)\hat{a}_n^+\hat{U}(t)|0\rangle = \frac{\alpha_n^*(t)}{|\lambda(t)|^2} \equiv \varphi_n^*(t). \qquad (23\text{-}7)$$

As a result, equations (23-5) take the form

$$i\dot{\varphi}_n = \Delta\varphi_n + \mathcal{M}(\varphi_{n+1} + \varphi_{n-1}) + \mathcal{M}'(\varphi_{n+1}^* + \varphi_{n-1}^*) + \\ + \mathcal{L}(|\varphi_{n+1}|^2 + |\varphi_{n-1}|^2)\varphi_n, \qquad (23\text{-}8)$$

where φ_n is the usual classical function having the meaning of the amplitude of probability that at the moment t the excitation is in the n-th lattice point.

Proceeding to the continuum limit we obtain finally

$$i\varphi_t = (\Delta + 2\mathcal{M})\varphi + \mathcal{M}\varphi_{xx} + 2\mathcal{M}'\varphi_x^* + 2\mathcal{L}|\varphi|^2\varphi. \qquad (23\text{-}9)$$

Analyzing the form of equation (23-9) we can conclude that the equation represents a modified variant of the nonlinear Schrödinger equation. In the linear approximation ($\mathcal{L}$=0) equation (23-9) takes the form

$$i\varphi_t = (\Delta + 2\mathcal{M})\varphi + \mathcal{M}\varphi_{xx} + 2\mathcal{M}'\varphi_x^*, \qquad (23\text{-}10)$$

and its solutions are usual plane waves

$$\varphi = \varphi_0 \exp[i(kx - \omega_k t)], \qquad (23\text{-}11)$$

with frequencies determined by the equation

$$\omega_\kappa = \{ [\Delta + 2\mathcal{M}(1-\kappa^2/2)]^2 + [2\mathcal{M}'\kappa]^2 \}^{1/2}. \tag{23-12}$$

In the Haitler-London approximation (M′=0), equation (23-9) transforms to

$$i\varphi_t = (\Delta + 2\mathcal{M})\varphi + \mathcal{M}\varphi_{xx} + 2\mathcal{L}|\varphi|^2\varphi. \tag{23-13}$$

It can be easily shown that (23-13) is completely equivalent to the nonlinear Schrödinger equation. Indeed, by the transformation

$$\varphi \to \tilde{\varphi} = \varphi \cdot \exp[i(\Delta + 2\mathcal{M})t], \quad x \to \tilde{x} = x/\sqrt{|\mathcal{M}|}, \quad (\mathcal{M} < 0), \tag{23-14}$$

equation (23-13) is reduced to the canonical form (18-1). Thus, besides usual solutions having the form of the plane waves, equation (23-13) has the soliton solution

$$\varphi = \frac{v^2 - 2uv \cdot \exp[i(\Delta+2\mathcal{M})t] \cdot \exp[i(v/2)(x/\sqrt{|\mathcal{M}|} - ut)]}{ch\{[(v^2-2uv)^{1/2}/2](x/\sqrt{|\mathcal{M}|} - vt)}, \tag{23-15}$$

A schematic picture of the solution is shown in Fig. 25.

The problems of the stability of the solutions are not considered here. We only note that it is well established that the soliton solutions of type (23-15) are more stable than the linear wave solutions of type (23-11) (phonons). Therefore, when discussing the ways of mathematical description of the process of the energy transmission in alpha-helical proteins the preference is given to the solitons. According to the hypothesis energy $\mathcal{E} \cong$ 0,43 eV, released as a result of the hydrolysis of ATP, can excite the intrapeptide oscillation Amid-1, and then this energy can be transmitted along the alpha-helical chain by soliton waves [63].

The accurate solution of the complete problem (23-9) could

shed an additional light on the question, but this solution has not been found yet. At present only the solution for the carrier wave Θ has been obtained [64, 65]. But this is not enough to make conclusions about the properties of the solutions of the complete problem.

§ 24. Transmission of nonlinear waves in alpha-helical proteins.

Let us consider now what changes in the results of § 23 will appear if we use some more complex variant of the physical model of alpha-helical protein, namely the model of Davydov and Suprun. A schematic picture of the model is shown in Fig. 17, and the model hamiltonian is given by formula (12-1). As in the previous case let us add corresponding quartic terms to the hamiltonian. As a result we obtain

$$\hat{H} = \sum_n \sum_{\alpha=1}^{3} \Delta \hat{P}^{+}_{n,\alpha} \hat{P}_{n,\alpha} + \sum_n \sum_{\alpha=1}^{3} M^{\alpha,\alpha}_{n,n+1} (\hat{P}^{+}_{n,\alpha} \hat{P}_{n+1,\alpha} + \hat{P}^{+}_{n+1,\alpha} \hat{P}_{n,\alpha}) +$$
$$+ \sum_n \sum_{\alpha=1}^{3} M'^{\alpha,\alpha}_{n,n+1} (\hat{P}^{+}_{n,\alpha} \hat{P}^{+}_{n+1,\alpha} + \hat{P}_{n,\alpha} \hat{P}_{n+1,\alpha}) + \sum_n \sum_{\alpha=1}^{3} Z^{\alpha,\alpha}_{n,n+1} \hat{N}_{n,\alpha} \hat{N}_{n+1,\alpha} +$$
$$+ \sum_n \{Q^{1,2}_{n,n} (\hat{P}^{+}_{n,1} \hat{P}_{n,2} + \hat{P}^{+}_{n,2} \hat{P}_{n,1}) + Q^{2,3}_{n,n} (\hat{P}^{+}_{n,2} \hat{P}_{n,3} + \hat{P}^{+}_{n,3} \hat{P}_{n,2}) +$$
$$+ Q^{3,1}_{n,n+1} (\hat{P}^{+}_{n,3} \hat{P}_{n+1,1} + \hat{P}^{+}_{n+1,1} \hat{P}_{n,3})\} + \sum_n \{Q'^{1,2}_{n,n} (\hat{P}^{+}_{n,1} \hat{P}^{+}_{n,2} + \hat{P}_{n,1} \hat{P}_{n,2}) + \tag{24-1}$$
$$+ Q'^{2,3}_{n,n} (\hat{P}^{+}_{n,2} \hat{P}^{+}_{n,3} + \hat{P}_{n,2} \hat{P}_{n,3}) + Q'^{3,1}_{n,n+1} (\hat{P}^{+}_{n,3} \hat{P}^{+}_{n+1,1} + \hat{P}_{n,3} \hat{P}_{n+1,1})\} +$$
$$+ \sum_n \{\mathcal{P}^{1,2}_{n,n} \hat{N}_{n,1} \hat{N}_{n,2} + \mathcal{P}^{2,3}_{n,n} \hat{N}_{n,2} \hat{N}_{n,3} + \mathcal{P}^{3,1}_{n,n} \hat{N}_{n,3} \hat{N}_{n+1,1}\}.$$

Using the rules of transformation from the Pauli-operators $\hat{P}^{+}_{n,\alpha}$, $\hat{P}_{n,\alpha}$ to the operators of Fermi $\hat{a}^{+}_{n,\alpha}$, $\hat{a}_{n,\alpha}$ described in §6, let us rewrite hamiltonian (24-1) in terms of Fermi-operators

$$\hat{H} = \sum_n \sum_{\alpha=1}^{3} \Delta \hat{a}^{+}_{n,\alpha} \hat{a}_{n,\alpha} + \sum_n M \{(1-2\hat{N}_{n,3})(1-2\hat{N}_{n,2}) (\hat{a}^{+}_{n,1} \hat{a}_{n+1,1} + \hat{a}^{+}_{n+1,1} \hat{a}_{n,1}) +$$
$$+ (1-2\hat{N}_{n+1,1})(1-2\hat{N}_{n,3})(\hat{a}^{+}_{n,2} \hat{a}_{n+1,2} + \hat{a}^{+}_{n+1,2} \hat{a}_{n,2}) + (1-2\hat{N}_{n+1,2})(1-2\hat{N}_{n,3}) \cdot$$
$$\cdot (\hat{a}^{+}_{n,3} \hat{a}_{n+1,3} + \hat{a}^{+}_{n+1,3} \hat{a}_{n,3})\} + \sum_n M' \{(1-2\hat{N}_{n,3})(1-2\hat{N}_{n,2})(\hat{a}^{+}_{n,1} \hat{a}^{+}_{n+1,1} -$$

$$-\hat{a}_{n,1}\hat{a}_{n+1,1}) + (1-2\hat{N}_{n+1,1})(1-2\hat{N}_{n,3})(\hat{a}^{+}_{n,2}\hat{a}^{+}_{n+1,2} - \hat{a}_{n,2}\hat{a}_{n+1,2}) + \quad (24\text{-}2)$$

$$+ (1-2\hat{N}_{n+1,2})(1-2\hat{N}_{n+1,1})(\hat{a}^{+}_{n,3}\hat{a}^{+}_{n+1,3} - \hat{a}_{n,3}\hat{a}_{n+1,3})\} +$$

$$+\sum_n \sum_{\alpha=1}^{3} \mathcal{Z}\hat{N}_{n\alpha}\hat{N}_{n+1,\alpha} + \sum_n Q\{(\hat{a}^{+}_{n,1}\hat{a}_{n,2} + \hat{a}^{+}_{n,2}\hat{a}_{n,1}) + (\hat{a}^{+}_{n,2}\hat{a}_{n,3} + \hat{a}^{+}_{n,3}\hat{a}_{n,2}) + (\hat{a}^{+}_{n,3}\hat{a}_{n+1,1} +$$

$$+ \hat{a}^{+}_{n+1,1}\hat{a}_{n,3})\} + Q'\{\hat{a}^{+}_{n,1}\hat{a}^{+}_{n,2} - \hat{a}_{n,1}\hat{a}_{n,2}) + (\hat{a}^{+}_{n,2}\hat{a}^{+}_{n,3} - \hat{a}_{n,2}\hat{a}_{n,3}) + (\hat{a}^{+}_{n,3}\hat{a}^{+}_{n+1,1} -$$

$$- \hat{a}_{n,3}\hat{a}_{n+1,1})\} + \sum_n \mathcal{P}\{\hat{N}_{n,1}\hat{N}_{n,2} + \hat{N}_{n,2}\hat{N}_{n,3} + \hat{N}_{n,3}\hat{N}_{n+1,1}\}.$$

To avoid long calculations, let us confine ourselves to a brief description of the plan of calculation.

1) Using hamiltonian (24-2) and formula (7-9) we obtain the equations of motion for the quantum operators $\hat{a}_{n,1}$, $\hat{a}_{n,2}$, $\hat{a}_{n,3}$ in the representation of Heisenberg.

2) With the help of $\hat{a} \to \Psi$ transformation and approximate procedure of uncoupling of tertiary terms we rewrite these equations in terms of usual classical functions $\varphi_{n,1}(t)$, $\varphi_{n,2}(t)$, $\varphi_{n,3}(t)$ ($\varphi_{n,\alpha}(t) \equiv \langle 0|\hat{U}^{+}(t)\hat{a}_{n,\alpha}\hat{U}(t)|0\rangle$, $\alpha=1,2,3$).

3) Passing to the continuum limit we obtain the unknown equations for the functions $\varphi_1(x,t)$, $\varphi_2(x,t)$, $\varphi_3(x,t)$ ($\varphi_\alpha(x,t) = \lim_{\text{discr.}\to\text{contin.}} \varphi_{n,\alpha}(t)$, $\alpha=1,2,3$) in the form

$$
\begin{aligned}
i\varphi_{1_t} &= (\Delta+2\mathcal{M})\varphi_1 + \mathcal{M}\varphi_{1_{xx}} + 2\mathcal{Z}|\varphi_1|^2\varphi_1 + (\mathcal{P}-8\mathcal{M})(|\varphi_2|^2+|\varphi_3|^2)\varphi_1 + \\
&\quad + Q\left[(\varphi_2+\varphi_3) + \tfrac{1}{3}(\varphi_{2_x}-\varphi_{3_x}) + \tfrac{1}{18}(\varphi_{2_{xx}}+\varphi_{3_{xx}})\right], \\
i\varphi_{2_t} &= (\Delta+2\mathcal{M})\varphi_2 + \mathcal{M}\varphi_{2_{xx}} + 2\mathcal{Z}|\varphi_2|^2\varphi_2 + (\mathcal{P}-8\mathcal{M})(|\varphi_3|^2+|\varphi_1|^2)\varphi_2 + \qquad (24\text{-}3) \\
&\quad + Q\left[(\varphi_3+\varphi_1) + \tfrac{1}{3}(\varphi_{3_x}-\varphi_{1_x}) + \tfrac{1}{18}(\varphi_{3_{xx}}+\varphi_{1_{xx}})\right], \\
i\varphi_{3_t} &= (\Delta+2\mathcal{M})\varphi_3 + \mathcal{M}\varphi_{3_{xx}} + 2\mathcal{Z}|\varphi_3|^2\varphi_3 + (\mathcal{P}-8\mathcal{M})(|\varphi_1|^2+|\varphi_2|^2)\varphi_3 + \\
&\quad + Q\left[(\varphi_1+\varphi_2) + \tfrac{1}{3}(\varphi_{1_x}-\varphi_{2_x}) + \tfrac{1}{18}(\varphi_{1_{xx}}+\varphi_{2_{xx}})\right].
\end{aligned}
$$

For simplicity we used here the Heitler-London approximation, when $M'=Q'=0$.

In the linear approximation equations (24-3) have solutions in the form of usual plane waves

$$\varphi_\alpha = \varphi_{0\alpha}\exp[i(kx-\omega_k t)] \qquad (24\text{-}4)$$

with dispersion relation with three branches of frequencies

$$\omega_{1_K} = \Delta + 2\mathcal{M}\left(1-\frac{K^2}{2}\right) + 2Q\left(1-\frac{K^2}{18}\right),$$

$$\omega_{2,3_K} = \Delta + 2\mathcal{M}\left(1-\frac{K^2}{2}\right) + 2Q\left(1-\frac{(K \pm 2\pi)^2}{18}\right). \qquad (24\text{-}5)$$

In the general case the nonlinear equations (24-3) have both the plane wave solutions and the soliton-like solutions. However, in contrast to the problem considered in §23, three types of soliton solutions are possible. The first one is determined by the formula

$$\varphi_1^{(1)} = \varphi_2^{(1)} = \varphi_3^{(1)} = \frac{A_1}{ch\,\nu_1 (x - v_1 t)} \exp i(Kx - \omega_1 t) \qquad (24\text{-}6)$$

Inserting (24-6) into (24-3) we can find the quantities v_1, ν_1 and ω_1

$$v_1 = -2(\mathcal{M} + Q/9)K,$$

$$(24\text{-}7)$$

$$\nu_1^2 = A_1^2(\mathcal{X} + \mathcal{P} - 8\mathcal{M})/(\mathcal{M} + Q/9), \qquad (24\text{-}8)$$

$$\omega_1 = \Delta + 2\mathcal{M}(1 - K^2/2) + 2Q(1 - K^2/18) + (\mathcal{X} + \mathcal{P} - 8\mathcal{M})A_1^2. \qquad (24\text{-}9)$$

Here A_1 is the constant determined by the conditions of normalization. The solution (24-6) is presented in Fig. 33a. It can be considered as a degenerated one, because three modulated waves ($\varphi_1^{(1)}$, $\varphi_2^{(1)}$ and $\varphi_3^{(1)}$) move in phase. The envelope of these waves has the form of a secant and moves with the velocity v_1.

The second (j=2) and the third (j=3) types of the soliton solutions have the form

$$\varphi_1^{(j)} = \frac{A_j}{ch\,\nu_j(x - v_j t)} \exp i(\kappa x - \omega_j t),$$

$$\varphi_2^{(j)} = \frac{A_j \exp(i\theta_j)}{ch\,\nu_j(x - v_j t)} \exp i(\kappa x - \omega_j t), \tag{24-10}$$

$$\varphi_3^{(j)} = \frac{A_j \exp(-i\theta_j)}{ch\,\nu_j(x - v_j t)} \exp i(\kappa x - \omega_j t), \qquad j = 2,3.$$

The index j denotes here the type of the solutions; θ_j is the phase displacement which is equal to

$$\theta_j = \begin{cases} 2\pi/3, & \text{if } j = 2, \\ 4\pi/3, & \text{if } j = 3. \end{cases} \tag{24-11}$$

If we insert (24-10) into (24-3) we can find the quantities v_j, ν_j and ω_j

$$v_{2,3} = -2\,[\mathcal{M} + (Q/9)\cos(2\pi/3)]\,\kappa \pm (2/3)Q\sin(2\pi/3), \tag{24-12}$$

$$\nu_{2,3}^2 = A_{2,3}^2\,(\mathcal{L} + \mathcal{P} - 8\mathcal{M})/[\mathcal{M} + (Q/9)\cos(2\pi/3)], \tag{24-13}$$

$$\omega_{2,3} = \Delta + 2\mathcal{M}(1 - \kappa^2/2) + 2Q(1 - \kappa^2/18)\cos(2\pi/3) \mp \tag{24-14}$$

$$\mp (2\kappa/3)Q\sin(2\pi/3) + (\mathcal{L} + \mathcal{P} - 8\mathcal{M})A_{2,3}^2.$$

The constants A_2 and A_3 are determined by the conditions of normalization. The solutions (24-10) are shown in Fig. 33b, c. In contrast to the solutions of the first type (Fig. 33a) the solutions of the second and third types are three modulated waves moving out of phase by $2\pi/3$ (when j=2) and by $4\pi/3$ (when j=3).

It is easy to check the results of the calculations carried out above if we compare the frequencies ω_1, ω_2 and ω_3 with those obtained in 12 for the linear discrete problem. Indeed, let us expand formulas (12-10) in powers of k. When the wave number (k) is small we can confine the expansion to quadratic terms

$$E_1 = \Delta + 2\mathcal{M}(1-k^2/2) + 2Q(1-k^2/18), \qquad (24\text{-}15)$$

$$E_{2,3} = \Delta + 2\mathcal{M}(1-k^2/2) + 2Q(1-k^2/18)\cos(2\pi/3) \mp \mp Q(2k/3)\sin(2\pi/3). \qquad (24\text{-}16)$$

Because we agreed earlier that $\hbar=1$, then $E_1=\omega_1$, $E_{2,3}=\omega_{2,3}$, and formulas (24-15) and (24-16) coincide with corresponding formulas (24-9) and (24-14) to an accuracy of only one term

$$(\mathcal{L}+\mathcal{P}-8\mathcal{M})A_j^2 \qquad (24\text{-}17)$$

the appearance of which is explained by the nonlinear effects.

§25. Transmission of nonlinear conformational waves in DNA. The model of Englander.

This paragraph is devoted to a new and debatable problem in studying DNA. A start was made by Englander and coauthors [66] in 1980. They had some difficulties when they tried to interpret experimental data on hydrogen-deuterium (and hydrogen-tritium) exchange. To explain the data they suggested that an opening of the pairs of nitrous bases occurs in DNA. To describe the process of opening, Englander and coauthors used the nonlinear sine-Gordon model. Why did they use this model? The answer is rather simple. The authors noticed a similarity between the rotational motions of the bases in DNA and the torsional oscillations of the pendula in the mechanical analog of the sG equation. To make this similarity more clear for the reader we present here a schematic picture of the model of DNA and the mechanical analog of the sG equation (see Fig. 34). It is now enough just to put in correspondence the pendula with the bases of one of the DNA strands, the springs with the sugar-phosphate basis, and the external gravitational field with the field induced by the second strand of DNA.

Using this analogy we can easily construct the equations describing the rotational oscillations of nitrous bases of one of the polynucleotide chains of DNA

$$I\ddot{\varphi}_n = K_0(\varphi_{n+1} - 2\varphi_n + \varphi_{n-1}) - V_0 \sin\varphi_n . \qquad (25\text{-}1)$$

Here φ_n is the angular displacement of the n-th base from its equilibrium position; I is the moment of inertia of the base; K_0

is the rigidity of the sugar-phosphate backbone; $V_0 \sin \varphi_n$ is the force acting upon the n-th base.

Proceeding to the continuum approximation by the rules described in § 17, we obtain

$$I\varphi_{tt} = K_0 a^2 \varphi_{zz} - V_0 \sin\varphi. \tag{25-2}$$

By substituting the variables

$$t \to t' = \sqrt{V_0/I}\; t, \quad z \to = \sqrt{V_0/K_0 a^2}\; z \tag{25-3}$$

equation (25-2) is reduced to the canonical form (19-1), and hence, it has one-soliton solution in the form of kink (or antikink)

$$\varphi = 4 \operatorname{arctg} \{\exp \pm [\gamma (z - z_0 - vt)/d]\}, \tag{25-4}$$

where $d = (K_0 a^2 / V_0)^{1/2}$. A schematic picture of the solution (25-4) is presented in Fig. 27.

Interpretation of kink solution (25-4) was discussed briefly at the beginning of this paragraph. A more detailed consideration of the question was carried out by Sobell [67]. According to his opinion, kink (25-4) describes a movable boundary between two phases, the bases in one of the phases being open and those in the other phase being closed. Sobell suggested also that the solution consisting of a pair of coupled solitons (kink and antikink) is possible. This two-soliton solution describes a movable local range with open base pairs inside and closed ones outside (Fig. 35). These regions are usually named "open states" and their existence is associated with the effects of the so-called "breathing" of DNA.

§26. Other nonlinear models simulating the transmission.

About the model of Englander and coauthors [66] we can say that it is simple and elegant. However, we should note also that the model is very far from the real DNA. It does not take into account many details of the structure and motions in DNA. In this paragraph we give a brief survey of other nonlinear models where the authors tried to improve the mathematical description of the DNA conformational dynamics.

1. G-L-F model.

The Klein-Gordon equation (20-5)

$$I\varphi_{tt} - K_0 a^2 \varphi_{zz} + \frac{\partial V}{\partial \varphi} = 0, \tag{26-1}$$

where $V(\varphi)$ is interpreted as a potential function describing interaction between the nitrous bases inside the pairs, can be considered as the simplest generalization of the model of Englander and coauthors [66]. In their model the function $V(\varphi)$ is modeled by

$$V(\varphi) = V_0 (1 - \cos\varphi). \tag{26-2}$$

Another form of the function V

$$V = (al^2 - b)\, \Theta(l_1 - l) \tag{26-3}$$

was proposed in the works of Gochev, Lisy and Fedyanin [68-69] (we name their model the G-L-F model). Here l is the distance between the bases inside the pair; a, b, l_1 are constants; θ is the function of Heavyside. We recall that θ is determined by

$$\theta(x) = \begin{cases} 1, & \text{if } x > 0, \\ 0, & \text{if } x < 0. \end{cases} \qquad (26\text{-}4)$$

A schematic picture of the potential function V is shown in Fig. 36. Because l is directly proportional to $\sin\varphi$ the potential V is a function of the variable .

The model potential (26-3) represents an improved variant of the model of Englander and coauthors. It accurately takes into account the ground state of the model system and the possibility of hydrogen bonds to be broken when the pair being open. Gochev, Lisy and Fedyanin succeeded in finding the soliton solution of equation (26-1) with potential (26-3) and showed that the solution has the form of quasikink.

2. The model of Yomosa.

Yomosa [70] proposed to model the rotational motions of nitrous bases by the two equations

$$I\varphi_{1_{tt}} - K_0 a^2 \varphi_{1_{zz}} + A \sin\varphi_1 + B \sin\varphi_1 \cos\varphi_2 = 0,$$
$$I\varphi_{2_{tt}} - K_0 a^2 \varphi_{2_{zz}} + A \sin\varphi_2 + B \sin\varphi_2 \cos\varphi_1 = 0, \qquad (26\text{-}5)$$

where φ_1 and φ_2 are the angle displacements of the bases which belong to the first and the second DNA strands, respectively; A and B are constants.

Equations (26-5) are a more correct version of the model of Englander because they take into account the rotational motions of bases in both DNA chains. We can rewrite equations (26-5) in a more general form

$$I_1 \varphi_{1_{tt}} - K_1 a^2 \varphi_{1_{zz}} + \frac{\partial V}{\partial \varphi_1} = 0,$$

$$I_2 \varphi_{2_{tt}} - K_2 a^2 \varphi_{2_{zz}} + \frac{\partial V}{\partial \varphi_2} = 0. \qquad (26\text{-}6)$$

Here $V(\varphi_1, \varphi_2)$ is the potential function simulating interactions between the bases in pairs. The form of the potential function

$$V(\varphi_1, \varphi_2) = A(1-\cos\varphi_1) + A(1-\cos\varphi_2) + B(1-\cos\varphi_1 \cdot \cos\varphi_2) \qquad (26\text{-}7)$$

has been obtained by Yomosa by expanding it in series of Fourier and by taking into account the general symmetrical properties of the model system. Very closed equations have been proposed independently by Takeno and Homma [71, 72].

The general solutions of equations (26-5) have not been found yet. However, some particular solutions can be easily obtained. Indeed, when $\varphi_1 = -\varphi_2$ equations (26-5) reduce to the known sG equation, when $\varphi_1 = \varphi_2$ they reduce to the dsG equation. Both these equations have in turn the soliton solutions of kink- (and antikink-) types (see §19 and §20).

3. <u>G-L-F-Y model.</u>

Another form of the function V

$$V(\varphi_1, \varphi_2) = \bar{a}(1-\cos\varphi_1) + \bar{a}(1-\cos\varphi_2) - \bar{b}(1-\cos(\varphi_1+\varphi_2)) \qquad (26\text{-}8)$$

has been proposed in the works [62, 68, 69]. Here a and b are constants. (In the title of the item the first letters of the authors' names are used). The equations of motion, which correspond to potential function (26-8), are

$$I_1 \varphi_{1_{tt}} - K_1 a^2 \varphi_{1_{zz}} + \bar{a} \sin \varphi_1 - \bar{b} \sin(\varphi_1 + \varphi_2) = 0,$$
$$I_2 \varphi_{2_{tt}} - K_2 a^2 \varphi_{2_{zz}} + \bar{a} \sin \varphi_2 - \bar{b} \sin(\varphi_1 + \varphi_2) = 0. \qquad (26\text{-}9)$$

The general solution of these equations has not been found yet. Only the first integral and the particular soliton-like solutions having the form of kinks and antikinks have been obtained in the cases when $\varphi_1 = -\varphi_2$ or $\varphi_1 = \varphi_2$ [72].

4. The model of Zhang.

Models described in the items 2-3 differ from one another by the form of the potential function V. In the work of Zhang [73] one more variant of the function is proposed

$$V = \alpha(1 - \cos\varphi_1) + \alpha(1 - \cos\varphi_2) + \beta(1 - \cos\varphi_1 \cdot \cos\varphi_2) - \\ -\mu(1 - \cos(\varphi_1 + \varphi_2)). \qquad (26\text{-}10)$$

To obtain this form of the potential function V the author took into account the energy of hydrogen bonds, the energy of dipole-dipole interactions and the energy of dipole-induced interactions.

Probably, only the results of simulation of the internal DNA dynamics by computers will help us to elucidate which of the potential functions mentioned above is preferable.

The equations of motions corresponding to potential (26-10) have the form

$$
\begin{aligned}
&I\varphi_{1_{tt}} - K_0 a^2 \varphi_{1_{zz}} + \alpha \sin\varphi_1 + \beta \sin\varphi_1 \cos\varphi_2 - \mu \sin(\varphi_1+\varphi_2) = 0, \\
&I\varphi_{2_{tt}} - K_0 a^2 \varphi_{2_{zz}} + \alpha \sin\varphi_2 + \beta \sin\varphi_2 \cos\varphi_1 - \mu \sin(\varphi_1+\varphi_2) = 0.
\end{aligned}
\qquad (26\text{-}11)
$$

As in the previous two cases the general solution of problem (26-11) has not been found yet except for the two particular cases which were mentioned in the items 2-3.

5. The model of Volkov and that of Krumhansl and Alexander.

In this item we shall consider another group of nonlinear models of the DNA dynamics, where besides the rotational motions of nitrous bases some other internal motions are taken into account. So, in the model of Volkov [74, 75] the puckering motions of sugars are considered. Their mobility is described as a movement in the double-well potential V (Fig. 30a) with minima which corresponds to the endo- and exo-conformational states of sugars. The reader who carefully read § 20 can guess that in the continuum limit the dynamics of sugars should be described by the nonlinear equation ^{4} . Just this equation has been used by Volkov.

In the work of Krumhansl and Alexander [76] an attempt has been made to consider three types of internal motions in DNA, namely: the rotational oscillations of the nitrous bases, the puckering motions of the sugars and the longitudinal displacements of the nucleosides. The model equations proposed by the authors have the following form

$$
\begin{aligned}
&\Omega \Phi^{(i)}_{tt} - C - A\Phi^{(i)} + B[\Phi^{(i)}]^3 - K\Phi^{(i)}_{zz} - \chi_0 Z^{(i)}_{zz} + \chi_1 Z^{(i)}_{z} - \chi_2 \varphi^{(i)}_{zz} = 0, \qquad (26\text{-}12)\\
&\mathcal{M} Z^{(i)}_{tt} - \kappa Z^{(i)}_{zz} - \chi_0 \Phi^{(i)}_{zz} - \chi_1 \Phi^{(i)}_{z} = 0, \\
&I\varphi^{(i)}_{tt} - \alpha \varphi^{(i)}_{zz} - \chi_2 \Phi^{(i)}_{zz} = 0, \qquad i = 1, 2.
\end{aligned}
$$

Here the functions $\mathcal{P}^{(i)}(z,t)$, $Z^{(i)}(z,t)$ and $\varphi^{(i)}(z,t)$ describe the sugar puckering, the nucleoside displacements and the rotations of bases, respectively. The index i denotes the number of the chain; A, B, C, K, k, Ω, I, M, α, χ_0, χ_1 and χ_2 are the constant parameters of the model system. Krumhansl and Alexander succeeded in finding the soliton-like solutions for the functions $\mathcal{P}^{(i)}(z,t)$ and the derivatives $Z_z^{(i)}(z,t)$ and $\varphi_z^{(i)}(z,t)$ (i=1,2)

$$\mathcal{P}^{(1)} = \mathcal{P}^{(2)} = \pm \sqrt{A_1/B}\; th[\mu(z - vt)],$$

$$Z_z^{(1),(2)} = \sqrt{A_1/B}\,\{\mu \chi_0\, sch^2[\mu(z - vt)] \pm th[\mu(z - vt)]\}, \qquad (26\text{-}13)$$

$$\varphi_z^{(1)} = \varphi_z^{(2)} = \sqrt{A_1/B}\;\mu\,\chi_2\; sch^2[\mu(z - vt)],$$

where v is the velocity, $\mu = \sqrt{A_1/2K_1}$, $A_1 = A - \chi_1^2/(k - Mv^2)$,

$$K_1 = K - \Omega v^2 - \chi_0^2/(k - Mv^2) - \chi_2^2/(\alpha - Iv^2).$$

A schematic picture of the solutions of the equations of Krumhansl and Alexander is presented in Fig. 37. The solution for the function $\mathcal{P}$ has the form of kink, and the solutions for the derivatives Z_z, φ_z have the form which is very similar to the form of localized traveling solitary waves obtained in §18 for the envelope of the nonlinear Schrödinger equation. It is obvious that the function Z and φ by themselves will have the form of kinks.

§27. Effects of dissipation and inhomogeneity.

The dynamical models of biopolymers, considered in §23-26 have two deficiencies: they describe only conservative and homogeneous systems. In the real systems both these conditions are not valid: the transmission of conformational waves in DNA or in proteins is usually accompanied by energy dissipation, and the primary structure of the polypeptide and polynucleotide chains is substantially inhomogeneous, namely it is characterized by the definite sequence of the amino acid residues (in proteins) and that of the nitrous bases (in DNA) (see §1,2).

In this paragraph we shall illustrate the method by which the effects of dissipation and inhomogeneity can be taken into account in the framework of the linear theory of perturbations. As an example let us consider the problem of the transmission of the nonlinear conformational waves in DNA.

As an ideal (or unperturbed) problem let us take the equation of Englander and coauthors (25-2)

$$\varphi_{tt} - c_o^2 \varphi_{zz} + \omega_o^2 \sin\varphi = 0, \qquad (27-1)$$

which has one-soliton solution in the form

$$\varphi^{(o)}(z,t) = 4\,\mathrm{arctg}\,\{\exp[(z - vt)\gamma/d]\} \qquad (27-2)$$

Here $d=c_o/\omega_o$, $c_o=(K_o a_o^2/I)^{1/2}$, $\omega_o=(V_o/I)^{1/2}$.

First, we shall consider the effects of dissipation. We can assume that these effects are caused by two reasons: by the existence of some internal friction and by interaction with the external surroundings. The simplest way of taking into account these two effects is to add to (27-1) two additional terms: the term $\beta\varphi_t$, which takes into account the internal friction, and the term $f_o(z,t)$ describing the action of the external fields. As a result the complete (or perturbed) problem will take the following form

$$\varphi_{tt} - c_o^2 \varphi_{zz} + \beta\varphi_t + \omega_o^2 \sin\varphi = f_o(z,t), \qquad (27-3)$$

where f_o =force/I, β=viscosity/I.

To solve the problem (27-3) let us use the results obtained in § 21. If we compare formulas (27-3) and (21-2), it becomes obvious that problem (27-3) is a particular case of problem (21-2) with perturbation V

$$V = f_o(z,t) - \beta\varphi_t . \qquad (27-4)$$

Therefore, we can use the results obtained in § 21, and consider the change of the velocity of the soliton wave due to the effects of dissipation. For this purpose it is necessary to give a concrete form for the function $f_o(z,t)$, then to solve equation (21-19) in the coefficients and then to use a simple relation between the derivative $\varphi_{b\,t'}(t')$ and the velocity, which is given by (21-31).

Let us illustrate this scheme of calculation for a simple case when $f_o = -J_o$ =Const. In this case equation (21-19) takes the form

$$\Psi_{b_{t't'}} = \frac{8\beta\gamma}{d} - 2\pi J_0 . \qquad (27\text{-}5)$$

Integrating (27-5) and assuming that $\Psi_{b_{t'}}(t'=0)=0$, we find

$$\Psi_{b_{t'}} = (8\beta\mu\gamma/d - 2\pi J_0)\, t', \qquad (27\text{-}6)$$

whence we obtain the formula of the velocity of soliton wave

$$\Delta v \cong -\frac{d}{8}\,\varphi_{b_{t'}} = (J_0 \pi d/4 - 8\mu\gamma)\, t'. \qquad (27\text{-}7)$$

determined in the laboratory coordinate system (t', z'). It follows from this formula that the movement of the soliton wave is accelerated due to the constant "pumping" force $f_0 = -J_0$ and decelerated due to the effects of the internal friction. We can give here the critical value

$$J_0^{crit.} = 4\beta\mu\gamma/\pi d, \qquad (27\text{-}8)$$

at which the action of the internal friction and "pumping" force becomes balanced.

Now let us turn to the second problem and consider the effect of inhomogeneity of the DNA structure. The most general method of taking into account the effect of inhomogeneity consists in substitution of the constant parameters of equation (27-1) by functions depending of the variable z

$$C_0 \rightarrow c(z), \qquad \omega_0 \rightarrow \omega(z). \tag{27-9}$$

However, we shall consider here a more simple case when only one of the two parameters, namely ω_0, is substituted by a function of z. In this case the perturbed problem takes the form

$$\varphi_{tt} - c_0^2 \varphi_{zz} + \omega_0^2 \sin\varphi + \lambda(z)\sin\varphi = 0, \tag{27-10}$$

where $\lambda(z) = \omega^2(z) - \omega_0^2$. By comparing (27-10) with equation (21-2) one can easily note that (27-10) is a particular case of problem (21-2) with the perturbation V

$$V = -\lambda(z)\sin\varphi. \tag{21-11}$$

Hence, we can use again the results obtained in §21. To illustrate this scheme of calculation let us calculate the change of the soliton wave velocity caused by inhomogeneity of the following type

$$\lambda(z) = \lambda_0\,\Theta(z - z_0), \tag{27-12}$$

where λ_0 is a constant that can be positive or negative depending on concrete situations; $\Theta(z-z_0)$ is the step function of Heavyside

$$\Theta(z - z_0) = \begin{cases} 1, & \text{if} \quad z > z_0, \\ 0, & \text{if} \quad z < z_0. \end{cases} \tag{27-13}$$

The model function (27-12) describes the boundary between two region, which is placed at the point $z=z_0$. A schematic picture of the function $\lambda(z)$ for the case when $\lambda_0>0$ is shown in Fig. 38a. It can be assumed, for example, that the region which is placed on the left of the boundary, is saturated by A–T pairs and the other region which is placed on the right, is saturated by G–C pairs (Fig. 38b).

To find the change of the velocity of solitary wave, let us use equation (21-19). In the considered case this equation can be written in the form

$$\Psi_{b_{t't'}} = 2\lambda_0 / ch^2[(vt' - z_0/\mu)/d]. \qquad (27\text{-}14)$$

Integrating (27-14) and assuming that $\Psi_{b_{t'}}(t'=-\infty)=0$ we find

$$\Psi_{b_{t'}} = (2\lambda_0 d/v)\{th[(vt' - z_0/\mu)/d] + 1\}, \qquad (27\text{-}15)$$

whence we find the velocity of the solitary wave (in the movable coordinate system)

$$\Delta v \cong -\frac{d}{8}\Psi_{b_{t'}} = -\frac{\lambda_0 d^2}{4v}\{1 + th[(vt' - z_0/\mu)/d]\}. \qquad (27\text{-}16)$$

This result shows that when crossing the boundary of two regions the velocity of the soliton wave increases or decreases depending on the sign (+ or −) of λ_0. For example, in the case shown in Fig. 38 ($\lambda_0>0$), the velocity of the soliton wave moving from the left to the right, decreases.

The result obtained can be easily extended to the case when we have some special standing out region of finite length. It can be, for example, a G-C block against an A-T background (Fig. 39). This problem differs from the previous one only by the existence of two boundaries (instead of one): the first is placed at the point $z=-a$ and the second at the point $z=a$. Hence, the model function $\lambda(z)$ is equal to

$$\lambda(z) = \lambda_0 \left[\Theta(z-a) - \Theta(z+a)\right], \quad (27-17)$$

and the change of the velocity calculated by the same scheme as in the previous case, is determined by

$$\Delta v = -\frac{\lambda_0 d^2}{4v} \left\{ th\left[(vt' - a/\mu)/d\right] - th\left[(vt' + a/\mu)/d\right] \right\}. \quad (27-18)$$

Formula (27-18) shows that when $\lambda_0 > 0$ the soliton decelerates when passing through the first (left) boundary and accelerates when passing through the second (right) one.

We can estimate the minimal value of the velocity of soliton wave ($v_{min.}$), which is necessary to surmount the energy barrier ($\lambda_0 > 0$) (Fig. 39) or the trap potential well ($\lambda_0 < 0$) (Fig. 40) and to continue the movement. For this purpose let us introduce the parameter E, which is equal to the barrier height (or to the well depth) and write the condition of the passage in the form

$$E_{kin.} \geqslant \Delta E, \quad (27-19)$$

where $E_{kin.}$ is the kinetic soliton energy determined by the formula

$$E_{kin.} = E_0 (\mu - 1). \quad (27-20)$$

Here E_0 is the rest energy. From (27-19) and (27-20) we find the unknown condition for the velocity

$$v^2 \geqslant v^2_{min.} \cong c_0^2 \frac{2\Delta E}{E_0} . \quad (27-21)$$

§28. Scattering of neutrons and light.

As it follows from the material of the previous paragraphs the theory gives an affirmative answer to the question concerning the possibility of transmission of nonlinear waves in biopolymers. And what about the experimental corroboration? Unfortunately, there is not any definite answer. In spite of the fact that most of the theoretical models of the nonlinear conformational dynamics just appeared as a result of the attempts to explain the contradictions in the experimental data available, the question about the validity of the idea of transmission of nonlinear waves in biopolymers still remains open, because no experiment which could give a convincing evidence has been carried out yet.

Theoreticians propose different ways of checking the theory. One of them is the scattering of neutrons. The experience stored by physicists dealing with the detection of solitons in quasi-one-dimensional magnetics (for example, in $CsNiF_3$), suggest to look just in this direction. Thus, the possibility of transmission of soliton waves in the magnetics was first predicted theoretically [77] and then it was confirmed experimentally [78, 79] by the method of the scattering of thermal neutrons. Another way is the scattering of infra-red light.

As a scatterer let us take a pattern consisting of DNA molecules which are organized in the way described in §13. We shall try to calculate contribution of thermal neutrons scattered by the nitrous bases of DNA, to the differential section of scattering. The reader has looked already through the method of neutron scattering and a scheme of calculation of the differential section of scattering. An important step in the scheme is the construction of the potential of interaction of the neutrons with the scatterer. In §13 this potential was modeled by operator $\hat{V}$ and defined by formula (13-5). In the case considered in this paragraph, the potential can be written in the form of a usual function $V(\vec{R})$

$$V(\vec{R}) = \frac{2\pi\hbar^2}{m_n} \sum_{n=1}^{N} a_n \,\delta(\vec{R} - \vec{R}_n), \qquad (28\text{-}1)$$

where $\vec{R}_n$ is the coordinate of the n-th nitrous base; a_n is the amplitude of the scattering by the n-th base ; $\vec{R}$ and m_n are the coordinate and mass of the neutron, respectively; $\hbar$ is the constant of Plank divided by 2π.

Let us denote the coordinates of the equilibrium positions of the bases by $\vec{R}^o_n$ and displacements from these positions by symbols $\vec{u}_n$ ($\vec{u}_n = \vec{R}_n - \vec{R}^o_n$). Then formula (28-1) can be rewritten in the form

$$V(\vec{R}) = \frac{2\pi\hbar^2}{m_n} \sum_{n=1}^{N} a_n \,\delta(\vec{R} - \vec{R}^o_n - \vec{u}_n). \qquad (28\text{-}2)$$

Using then the classical theory of scattering [80] and formula (28-2) we can easily obtain the differential section of scattering

$$\frac{d^2\sigma}{d\Omega\, dE'} = N\frac{\kappa'}{\kappa}\bar{a}^2 S^{coh.}(\vec{q},\omega) + N\frac{\kappa'}{\kappa}(\overline{a^2} - \bar{a}^2)\, S^{incoh.}(\vec{q},\omega), \qquad (28\text{-}3)$$

where

$$S^{coh.}(\vec{q},\omega) = \frac{1}{2\pi\hbar N}\sum_{n,n'} e^{-i\vec{q}(\vec{R}^o_n - \vec{R}^o_{n'})}\int_{-\infty}^{+\infty} dt\, e^{i\omega t} \langle e^{-i\vec{q}\vec{u}_n(t)}, e^{i\vec{q}\vec{u}_{n'}(0)}\rangle,$$

(28-4)

$$S^{incoh}(\vec{q},\omega)=\frac{1}{2\pi\hbar N}\sum_n\int_{-\infty}^{+\infty}dt\, e^{i\omega t}\langle e^{-i\vec{q}\vec{u}_n(t)}; e^{i\vec{q}\vec{u}_n(0)}\rangle. \qquad (28\text{-}5)$$

The first and the second terms in (28-3) describe the contribution of coherent and incoherent scattering, respectively; $\hbar\vec{k}$ and $\hbar\vec{k}'$ are the impulses of neutrons before and after the scattering; $\hbar\omega=(\hbar k'^2-\hbar k^2)/2m$ is the change of the neutron energy due to the scattering; $\vec{q}=(\vec{k}'-\vec{k})$; $\overline{\ldots}$ denotes averaging over orientations of spins and isotopes; $\langle\ldots\rangle$ denotes averaging over the positions of bases.

Expanding exponents in (28-4) and (28-5) we can rewrite the values $S^{coh.}$ and $S^{incoh.}$ in the form of the infinite sum of terms. The first term will describe the contribution given by elastic scattering, the other terms will describe inelastic scattering. For example, the second term will describe the contribution given by inelastic scattering leading to the creation or annihilation of only one soliton. The third term will describe the scattering leading to the creation or annihilation of two solitons, and so on. Approximations which are limited by the second, third and higher terms are named one-particle approximation, two-particle approximation, three-particle approximation and so on.

To illustrate the method let us calculate in the one-particle approximation the contribution given by coherent inelastic scattering

$$S^{coh.inel.}(\vec{q},\omega)=\frac{1}{2\pi\hbar N}e^{-2W_{\vec{q}}}\sum_{n,n'}e^{i\vec{q}(\vec{R}^o_n-\vec{R}^o_{n'})}\int_{-\infty}^{+\infty}e^{i\omega t}\langle\vec{q}\vec{u}_n(t);\vec{q}\vec{u}_{n'}(0)\rangle. \qquad (28\text{-}6)$$

Here $W_{\vec{q}}$ is the Debye-Waller factor. To calculate $S^{coh.\,inel.}(\vec{q},\omega)$ we need to find the values $\vec{R}_n^o$, $\vec{q}$ and $\vec{u}_n$. We can do it in a simple way, if we use the mechanical analog of the DNA model shown in Fig. 34b. It follows from this figure that

$$\vec{R}_n^o = \{R;\,0;\,nd\},\quad \vec{R}_n = \{R\cos\varphi_n;\,R\sin\varphi_n;\,nd\},$$

$$\vec{u}_n = \vec{R}_n - \vec{R}_n^o = \{R(\cos\varphi_n - 1);\,R\sin\varphi_n;\,0\}, \tag{28-7}$$

$$\vec{q} = \{q_x;\,q_y;\,q_z\},$$

where R is the length of pendula. Inserting (28-7) into (28-6) we find

$$S^{coh.\,inel.}(\vec{q},\omega) = S_{\parallel}(\vec{q},\omega) + S_{\perp}(\vec{q},\omega), \tag{28-8}$$

where $S_{\parallel}$ and $S_{\perp}$ are the longitudinal and transversal components of the dynamical form factor determined by the formulas

$$S_{\parallel}(\vec{q},\omega) = \frac{R^2 q_x^2}{2\pi\hbar N} e^{-2W_{\vec{q}}} \sum_{n,n'} e^{-iq_z a(n-n')} \int_{-\infty}^{+\infty} dt\, e^{i\omega t} \langle(1-\cos\varphi_n(t));\,(1-\cos\varphi_{n'}(0))\rangle, \tag{28-9}$$

$$S_{\perp}(\vec{q},\omega) = \frac{R^2 q_y^2}{2\pi\hbar N} e^{-2W_{\vec{q}}} \sum_{n,n'} e^{-iq_z a(n-n')} \int_{-\infty}^{+\infty} dt\, e^{i\omega t} \langle \sin\varphi_n(t);\,\sin\varphi_{n'}(0)\rangle. \tag{28-10}$$

In the continuum approximation ($z = an \to z$, $\varphi_n(t) \to \varphi(z,t)$) formulas (28-9) and (28-10) transform to

$$S_{\parallel}(\vec{q},\omega) = \frac{R^2 q_x^2}{2\pi\hbar N} e^{-2W_{\vec{q}}} \int_{-\infty}^{+\infty} dz' \int_{-\infty}^{+\infty} dz \int_{-\infty}^{+\infty} dt\, e^{-iq_z(z-z') + i\omega t} \times$$

$$\times \langle(1-\cos\varphi(z,t));\,(1-\cos\varphi(z',0))\rangle, \tag{28-11}$$

$$S_{\perp}(\vec{q},\omega) = \frac{R^2 q_y^2}{2\pi\hbar N} e^{-2W_{\vec{q}}} \int_{-\infty}^{+\infty} dz' \int_{-\infty}^{+\infty} dz \int_{-\infty}^{+\infty} dt \, e^{-iq_z(z-z')+i\omega t} \times$$
$$\times \langle \sin\varphi(z,t); \sin\varphi(z',0)\rangle . \tag{28-12}$$

Now it remains only to calculate the correlation functions

$$\langle (1-\cos\varphi(z,t)); (1-\cos\varphi(z',0))\rangle \ , \quad \langle \sin\varphi(z,t); \sin\varphi(z',0)\rangle .$$

If we take the model of Englander as a mathematical model describing the excited states of DNA, and if we construct the statistical ensemble of solitons as it was done in § 22, the problem of thermal neutron scattering by nitrous bases of DNA will reduce to the problem of neutron scattering by the ideal gas of solitons. We can simplify the calculation if we substitute the average $\langle\ldots\rangle$ by the approximate formula

$$\langle\ldots\rangle \cong n_s \langle\ldots\rangle_1 , \tag{28-13}$$

where n_s is the average number of solitons calculated in § 22, and the symbol $\langle\ldots\rangle_1$ denotes the averaging over the states of one (free) soliton

$$\langle\ldots\rangle_1 = \frac{\int dz \int dp_z \ldots \exp(-\mathcal{E}_s/k_B T)}{\int dz \int dp_z \exp(-\mathcal{E}_s/k_B T)} = \tag{28-14}$$
$$= 2m_0 c_0 \mathcal{K}_1(\mathcal{E}_0/k_B T) \int dp_z \ldots \exp(-\mathcal{E}_s/k_B T) .$$

We would like to remind the reader that $\mathcal{E}_0$ and m_0 are the rest energy and rest mass of soliton, which are determined in §22; $\mathcal{K}_1(x)$ is the function of Macdonald; c_0 is the constant parameter of the sG equation which is the basis of the model of Englander.

Inserting formulas (28-11)-(28-14) and soliton solution (25-4) into (28-8) we obtain

$$S^{coh.\,inel.}(\vec{q},\omega) \cong \frac{4R^2 a d \mu_0 (\mathcal{E}_0/2\pi\kappa_B T)^{1/2}}{\hbar c_0 q_z \mathcal{K}_1(\mathcal{E}_0/\kappa_B T)} e^{-2W_{\vec{q}}} e^{-\mathcal{E}_0(1+\mu_0)/\kappa_B T} \times \tag{28-15}$$

$$\times \{ q_x^2 [(\pi q_z d/\mu_0)/sh(\pi q_z d/2\mu_0)]^2 + q_x^2 [(\pi q_z d/\mu_0)/ch(\pi q_z d/2\mu_0)]^2 \},$$

where we used the notation $\mu_0 = (1 - w^2/q_z^2 c_0^2)^{-1/2}$.

In the case of low temperature ($T \ll \mathcal{E}_0/k_B$) and small velocities ($w/q \ll c_0$) we find a more simple variant of formula (28-15)

$$S^{coh.\,inel.}(\vec{q},\omega) \cong \frac{4R^2 a d \mathcal{E}_0}{\hbar c_0 q_z \pi \kappa_B T} e^{-2W_{\vec{q}}} \{ q_x^2 [(\pi q_z d)/sh(\pi q_z d/2)]^2 + \tag{28-16}$$

$$+ q_y^2 [(\pi q_z d)/ch(\pi q_z d/2)]^2 \} \exp(-\mathcal{E}_0/\kappa_B T) \exp(-m_0 \omega^2/2 q_z^2 \kappa_B T).$$

Note that the temperature corresponding to usual physiological conditions (or the room temperature) belongs to the range of so-called "low" temperatures, that is the condition $T \ll \mathcal{E}_0/k_B$ is valid. Indeed, if we use the values $\mathcal{E}_0 = 6$ kcal/mol, given in [66] and $k_B = 1,38 \times 10^{-23}$ j (^{o}K^{-1}) then $(\mathcal{E}_0/k_B) \cong 3 \times 10^3$ ^{o}K. The last value exceeds the room temperature by one order of magnitude. The second condition, $w/q_z \ll c_0$, means that formula (28-16) is valid for the soliton waves moving in DNA with velocities which are much smaller than the velocity of the acoustic waves $c_0 \cong 2 \times 10^3$ m/s [81].

Note that results (28-15) and (28-16) can be used also for interpretation of the experimental data on scattering of unfra-red light by DNA. Komarov, Fisher and Pecora [82, 83] showed that the spectral density of the scattered light is determined by the formula

$$I(\vec{q},\omega) = I_o \frac{\alpha^2 \omega^4}{2\pi c^4 \rho^2} \sin^2\bar{\mu} \sum_{n,n'} e^{i\vec{q}(\vec{R}_n^0 - \vec{R}_{n'}^0)} \int_{-\infty}^{+\infty} dt\, e^{i\omega t} \langle e^{-i\vec{q}\vec{u}_n(t)} ; e^{i\vec{q}\vec{u}_{n'}(0)} \rangle, \qquad (28\text{-}17)$$

where α is the polarizability of the bases; ω is the difference between the frequencies of the incident and scattered light; c is the velocity of light; ρ is the distance between the scattering system and the point of observation; the angle $\bar{\mu}$ and the intensity of incident light I_o are determined by the formulas

$$\cos\bar{\mu} = (\vec{E}_o \vec{\rho})/E_o \rho, \quad I_o = c|\vec{E}_o|^2/4\pi, \qquad (28\text{-}18)$$

where $\vec{E}_o$ is the vector of the amplitude of the wave of the incident light; the rest of the notations have the same sense as the notations in formulas (28-3)-(28-5). Comparing (28-17) and (28-4) we find

$$I(\vec{q},\omega) = I_o \frac{\alpha^2 \omega^4 \hbar N}{c^4 \rho^2} \sin^2\bar{\mu}\, S^{coh.}(\vec{q},\omega). \qquad (28\text{-}19)$$

It follows from this that results (25-15) and (28-16) can be applied to the problem of scattering of light. So, we can predict the existence of the central peak in the spectrum of scattering, the parameters of the peak (the integral intensity I and the peak width $\Delta\omega$) depending on the temperature T and on the wave vector $\vec{q}$

in some special way.

In the case of low temperature and small velocities of solitons these dependencies have the following form

$$I = \int S^{coh.\,inel.}(\vec{q},\omega)\,d\omega = (A(\vec{q})/\sqrt{\kappa_B T})\exp(-\mathcal{E}_0/\kappa_B T), \tag{28-20}$$

$$\Delta\omega = B(\vec{q})\sqrt{\kappa_B T}. \tag{28-21}$$

Here coefficients $A(\vec{q})$ and $B(\vec{q})$ do not depend on temperature, and they are determined by

$$A(\vec{q}) = \frac{4\sqrt{2}\,R^2 a d}{\hbar\sqrt{\pi\mathcal{E}_0}}\,e^{-2W_{\vec{q}}}\left\{q_x^2\left[\frac{\pi q_z d}{sh(\pi q_z d/2)}\right]^2 + q_y^2\left[\frac{\pi q_z d}{ch(\pi q_z d/2)}\right]^2\right\}, \tag{28-22}$$

$$B(\vec{q}) = \frac{q_z}{\sqrt{m_0}}. \tag{28-23}$$

The dependence of the parameters I and $\Delta\omega$ on the temperature is shown schematically in Fig. 41. When increasing the temperature, the intensity of the central peak increases exponentially and the peak width increases in proportion to the root of T.

§29. Effects of helicity.

In the previous paragraph we considered the scattering of thermal neutrons by solitons in DNA. To calculate the differential section of scattering a simple model proposed by Englander and coauthors was used. However, this model has many deficiencies. Some ways of improving the model have been discussed partly in §26-28. Here we shall consider one more possibility which is associated with taking into account the helical character of the DNA structure. Instead of the model of Englander having the mechanical analog in the form of a chain of pendula hanging on horizontal filament (Fig. 42a), we consider another model where the filament with pendula is assumed to be wound around the axis to form a helix (Fig. 42b), the points of hanging up being on the helix and the pivots of the pendula being directed to the axis. As a result the neighboring pendula are twisted relatively to one another by the angle $\varphi_0 = 36^o$. It is suggested also that the direction of "the external gravitational" field induced by the second chain changes from one pendulum to another by the angle $\varphi_0 = 36^o$.

We assume that the change of the model does not lead to substantial change of the form of the equations of motion. (An opposite case has been considered by Gaeta [84] and we shall discuss it briefly at the end of the paragraph). So, we can expect that there exist the soliton solutions of kink and antikink types.

Let us consider now the scattering of neutrons (and light) by these solitons and compare the results with those obtained previously for the case when the effects of the helicity were not taken into account. For this purpose let us write the vector coordinates $\vec{R}^{o}_{n}$, $\vec{R}_{n}$, $\vec{u}_{n}$ appearing in formula (28-6)

$$\vec{R}^{o}_{n}=\{(R-OO')\cos(2\pi n/10);\ -(R-OO')\sin(2\pi n/10);\ na\},$$
$$\vec{R}_{n}=\{(R\cos\varphi_{n}-OO')\cos(2\pi n/10)+(R\sin\varphi_{n})\sin(2\pi n/10);\ -(R\cos\varphi_{n}-OO')\sin(2\pi n/10)+$$
$$+(R\sin\varphi_{n})\cos(2\pi n/10);\ na\}, \tag{29-1}$$
$$\vec{u}_{n}=\vec{R}_{n}-\vec{R}^{o}_{n}=\{R\cos(2\pi n/10)(\cos\varphi_{n}-1)+R\sin(2\pi n/10)\sin\varphi_{n};\ -R\sin(2\pi n/10)(\cos\varphi_{n}-1)+$$
$$+R\cos(2\pi n/10)\sin\varphi_{n};\ 0\}.$$

Here OO' is the radius of the helix. Inserting the components of the vectors $\vec{R}^{o}_{n}$, $\vec{u}_{n}$ into (28-6) and following step by step the scheme of the calculation described in the previous paragraph, we find

$$S^{coh.\,inel.}(\vec{q},\omega)\cong\frac{R^{2}ad\sqrt{\mathcal{E}_{0}/2\pi\kappa_{B}T}}{\hbar c_{0}K_{1}(\mathcal{E}_{0}/\kappa_{B}T)}e^{-2W_{\vec{q}}}e^{-\mathcal{E}_{0}/\kappa_{B}T}(q_{x}^{2}+q_{y}^{2})\times$$
$$\times\{F_{+}(q_{z}+2\pi/10a)+F_{-}(q_{z}-2\pi/10a)\}, \tag{29-2}$$

where the notations

$$F_{\pm}(\xi)=\left\{\frac{\pi d\xi\,(1-\omega^{2}/c_{0}^{2}\xi^{2})^{1/2}}{sh\,[(\pi d\xi/2)(1-\omega^{2}/c_{0}^{2}\xi^{2})^{1/2}]}\pm\right.$$
$$\left.\pm\frac{\pi d\xi\,(1-\omega^{2}/c_{0}^{2}\xi^{2})^{1/2}}{ch\,[(\pi d\xi/2)(1-\omega^{2}/c_{0}^{2}\xi^{2})^{1/2}]}\right\}^{2}\cdot\frac{1}{\xi}(1-\omega^{2}/c_{0}^{2}\xi^{2})^{-1/2}\cdot\exp\{(-\mathcal{E}_{0}/\kappa_{B}T)(1-\omega^{2}/c_{0}^{2}\xi^{2})^{-1/2}\}. \tag{29-3}$$

are used.

For the range of low temperatures and small velocities we have

$$S^{coh.\,inel.}(\vec{q},\omega)\cong\frac{R^{2}ad\,\mathcal{E}_{0}}{\hbar c_{0}\pi\kappa_{B}T}e^{-2W_{\vec{q}}}(q_{x}^{2}+q_{y}^{2})\{f_{+}(q_{z}+2\pi/10a)+$$
$$+f_{-}(q_{z}-2\pi/10a)\}, \tag{29-4}$$

where

$$f_{\pm}(\xi) = \left[\frac{\pi d\xi}{sh(\pi d\xi/2)} \pm \frac{\pi d\xi}{ch(\pi d\xi/2)} \right]^2 \cdot \frac{1}{\xi} e^{-\mathcal{E}_0/\kappa_B T} \cdot e^{-m_0\omega^2/2\xi^2\kappa_B T} . \quad (29\text{-}5)$$

It follows from (29-2) and (29-4) that the central peak splits up (according to variable q) into two components, shifted towards each other by $4\pi/10a$ (Fig. 43). Splitting of such an order of magnitude ($2\pi/5a = 0,37\ \text{Å}^{-1}$) can be measured in the scattering of thermal neutrons obtained at present neutron sources.

The integral intensities and the line widths of the components are equal to

$$I_{\pm} = (A_{\pm}(\vec{q})/\kappa_B T)\exp(-\mathcal{E}_0/\kappa_B T) , \quad (29\text{-}6)$$

$$\Delta\omega_{\pm} = B_{\pm}(\vec{q})\sqrt{\kappa_B T} . \quad (29\text{-}7)$$

Here the coefficients $A_{\pm}$, $B_{\pm}$ do not depend on temperature and are determined by the formulas

$$A_{\pm}(\vec{q}) = \frac{\sqrt{2}R^2 a d}{\hbar\sqrt{\pi}\,\mathcal{E}_0} e^{-2W_{\vec{q}}} (q_x^2 + q_y^2) \left\{ \frac{\pi d(q_z \pm 2\pi/10a)}{sh[(q_z \pm 2\pi/10a)\pi d/2]} \pm \frac{\pi d(q_z \pm 2\pi/10a)}{ch[(q_z \pm 2\pi/10a)\pi d/2]} \right\}^2 , \quad (29\text{-}8)$$

$$B_{\pm}(\vec{q}) = (q_z \pm 2\pi/10a)/\sqrt{m_0} . \quad (29\text{-}9)$$

As follows from (29-6) and (29-7), the improving of the model of Englander due to taking into account the helical structure of the scatterer does not lead to essential changes of the parameters I and $\Delta\omega$. However, the reader can easily find these changes in the q-dependence of the parameters if he compares formulas (28-20)-(28-23) with (29-6)-(29-9).

In contrast to the model described above, Gaeta [84] proposed to take into account the changes in the equations of motions caused by the effects of helicity. He proposed a simple way of finding these changes. Besides the next nearest neighboring interactions between the pendula of the chain he includes into consideration the additional interactions between the pendula which are far apart along the chains but which are actually neighboring in space due to the helicity . This suggestion leads to the appearance of additional terms in the equation of motion. The reader can find the details in [84] and try to calculate the spectrum of neutron scattering using the model of Gaeta.

PART 6. CONCLUSIVE REMARKS. METHODS THAT HAVE NOT BEEN INCLUDED IN THE BOOK. HYPOTHESES AND PERSPECTIVES.

Here I would like to describe in a few words the methods that have not been included in the book and to discuss briefly the hypotheses and perspectives.

The diagram presented in Fig. 44 shows the fields of theoretical physics considered in this book. The choice of them as well as the choice of the methods and applications were determined by the interests of the author. Of course they do not exhaust the whole theme.

Let us list some of the questions which were not considered in the book. They are the methods of quantum chemistry and their applications to the calculations of the electron structure of the fragments of biomolecules, the methods of harmonic analysis (or lattice theory) and their applications to the calculations of the vibrational spectrum of biomolecules, the methods of the theory of phase transitions (including the Ising model) and their applications to studying conformational transitions in biomolecules (for example, the "helix-clew" transition in proteins or A $\rightarrow$ B transition in DNA).

The notion about the biomolecule as a "lifeless" physical system consisting of a number of atoms arranged in a definite way in space, is the bases of the physical approach to the investigation of biomolecules. In spite of that, many new physical ideas penetrate into "Sancta Sanctorum" of molecular biology, namely, into the problems of functioning of biomolecules. This concerns the ideas about the role of the lowest excited (electronic and vibrational) states of bimolecules in the energy transport processes, the ideas about the correlation between the conformation states of biopolymers and their functional activity, the hypothesis on regulatory role of the nonlinear solitary waves transmitting along the biopolymer chains and so on. It seems that the development of these ideas is very promising.

Let us note at least one more important moment concerning interconnection of theoretical physics and molecular biology. The attentive reader could note that we endeavored everywhere (where it was necessary) to stress the new and unsolved for the present mathematical problems appearing in the theory of biopolymers. The reverse connection (from molecular biology to mathematics) is also very perspective and "harbours" new interesting potentialities in the development of the mathematical basis of theoretical physics.

LIST OF NOTATIONS

$A(\vec{R})$ is the operator of the amplitude of scattering

E_k, $E(k)$ are the energy of elementary excitation

E_o is the energy of the ground state

$\mathcal{E}_o$ is the rest energy of a relativistic particle

G is the thermodynamic potential

$G(t,t')$ is the Green function

$G(E)$ is Fourier-transform of the Green function

$\mathcal{H}$, H are the hamiltonians

I_n is the n-th integral of motion; the moment of inertia of the n-th pendulum

$I(\vec{q},\omega)$ is the spectral density of the scattered light

I is the integral intensity

$J(\omega)$ is the spectral representation of the Green function

K_o is the rigidity

$\mathcal{L}$ is the lagrangian

N is the number of particles in a system

$\hat{P}^+$, $\hat{P}$ are the operators of Pauli

$\vec{R}$ is the vector of coordinates

R_i is the i-th amino acid residue

$S(\vec{q},\omega)$ is the dynamical form factor of scattering

T is the absolute temperature

$\hat{U}(t)$ is the operator of evolution

$\hat{V}$ is the operator of interaction of neutrons with a scattering center

V is a small correction (or perturbation) to unperturbed equation of hamiltonian

$V(q)$ is the potential energy

$W_{\vec{q}}$ is the Debye-Waller factor

Y is Young's modulus

$\theta(x)$ is the function of Heavyside

Ξ is the large statistical sum

Ψ is the wave function of state of a quantum system

Ω is the space angle

a is the lattice constant; the constant of equation

$\hat{a}^{+}$, $\hat{a}$ are the operators of Fermi

$\hat{b}^{+}$, $\hat{b}$ are the operators of Bose

c is the sound velocity

$\hbar$ is the constant of Planck

k_B is the constant of Boltzmann

k is the wave number; the constant of equation; the rigidity

m is the mass of a particle

m_n is the mass of neutron

m_o is the rest mass of a relativistic particle

n is the index numerating elementary cells of one-dimensional lattice

n_S is the density of solitons

p_i is the generalized impulse of the i-th particle

p is the ensemble of 3N generalized impulses of a system

$\hat{p}^{+}$, $\hat{p}$ are the operators of Pauli

q_i is the generalized coordinate of the i-th particle

q is the ensemble of 3N generalized coordinates of a system; the wave number

$\vec{q}$ is the wave vector

q_i is the velocity of the i-th particle

$\dot{q}$ is the ensemble of 3N velocities of the particles of a system

t is time

v is the velocity of nonlinear wave; the velocity of elementary excitation moving along the molecular chain

ω is the frequency

Δω is the width of spectral line

μ is the chemical potential

ρ is the linear density; the distance between the scatterer and the observation point

σ is the number of molecules in the lattice cell; the total section of scattering

$\frac{d^2\sigma}{dE_{\vec{p}'} d\Omega}$ is the differential section of scattering

τ is the time of transmission of the energy of excitation from one molecule to another (neighbouring) molecule

Table 1.

The most important amino acids occurring in proteins

$$R-CH(COOH)(NH_2)$$

N	Amino acid	Abbreviation	R-group
1.	Alanine	Ala	$-CH_3$
2.	Arginine	Arg	$-CH_2-CH_2-CH_2-NH-C(=NH)-NH_2$
3.	Asparagine	Asn	$-CH_2-C(=O)-NH_2$
4.	Asparaginic acid	Asp	$-CH_2-COOH$
5.	Cysteine	Cys	$-CH_2-SH$
6.	Glutamine	Gln	$-CH_2-CH_2-C(=O)-NH_2$
7.	Glutaminic acid	Glu	$-CH_2-CH_2-COOH$
8.	Glycine	Gly	$-H$
9.	Histidine	His	$-CH_2-C=CH-NH-CH=N-$ (C–N ring closure)
10.	Isoleucine	Ile	$-CH(CH_3)-CH_2-CH_3$

11.	Leucine	Leu	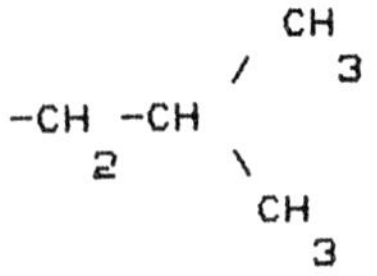
12.	Lysine	Lys	$-CH_2-CH_2-CH_2-CH_2-NH_2$
13.	Methionine	Met	$-CH_2-CH_2-S-CH_3$
14.	Phenyl alanine	Phe	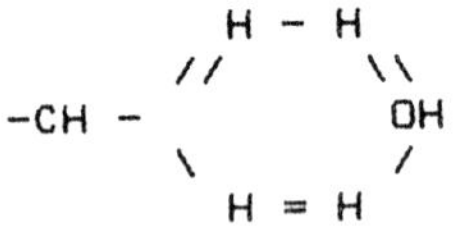
15.	Proline	Pro	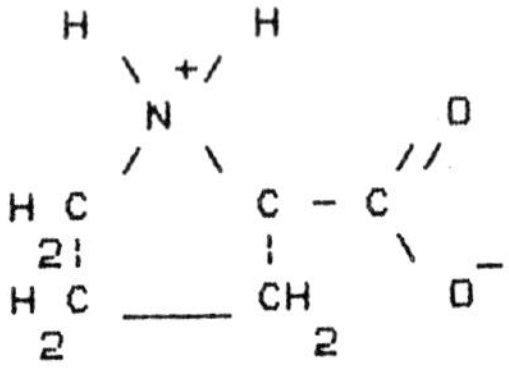
16.	Serine	Ser	$-CH_2OH$
17.	Threonine	Thr	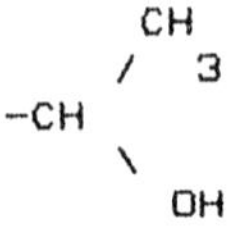
18.	Tryptophan	Try	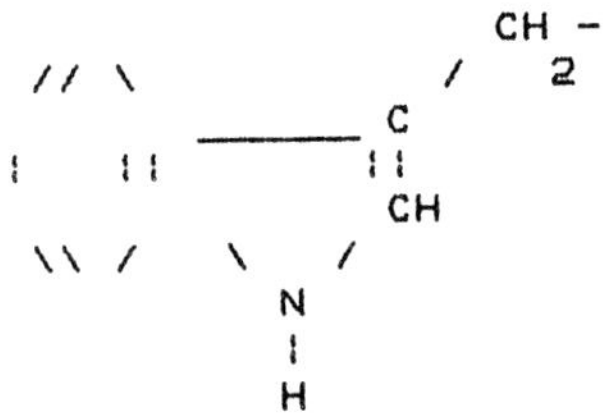
19.	Tyrosine	Tyr	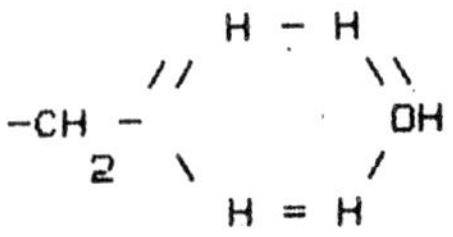
20.	Valine	Val	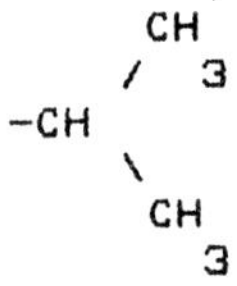

Table 2.

Sizes of different DNA molecules.

Organism	Number of bases	Length, Å	Diameter, Å
Viruses:			
Polyoma or SV-40	$5,1\cdot10^{3}$	$1,7\cdot10^{4}$	20
λ-phage	$4,86\cdot10^{4}$	$1,7\cdot10^{5}$	20
T2-phage	$1,66\cdot10^{5}$	$5,6\cdot10^{5}$	20
Viruses of cow-pox	$1,9\cdot10^{5}$	$6,5\cdot10^{5}$	20
Bacteria:			
Mycoplasma	$7,6\cdot10^{5}$	$2,6\cdot10^{6}$	20
E. coli	$4,0\cdot10^{6}$	$1,36\cdot10^{7}$	20
Eukaryotes:			
Yeast	$1,35\cdot10^{7}$	$4,6\cdot10^{7}$	20
Drosophila	$1,65\cdot10^{8}$	$5,6\cdot10^{8}$	20
Human	$2,9\cdot10^{9}$	$9,9\cdot10^{9}$	20

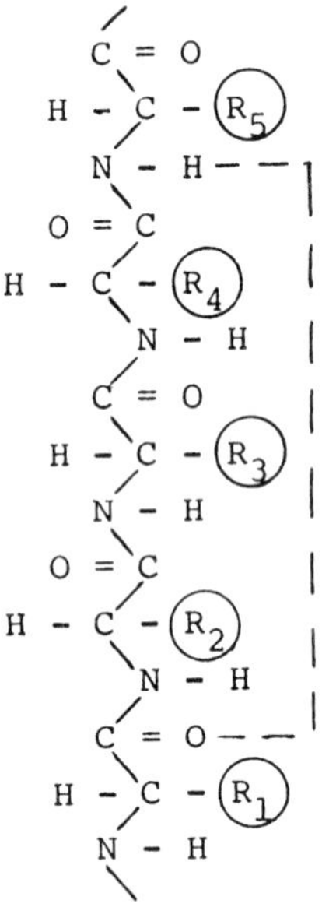

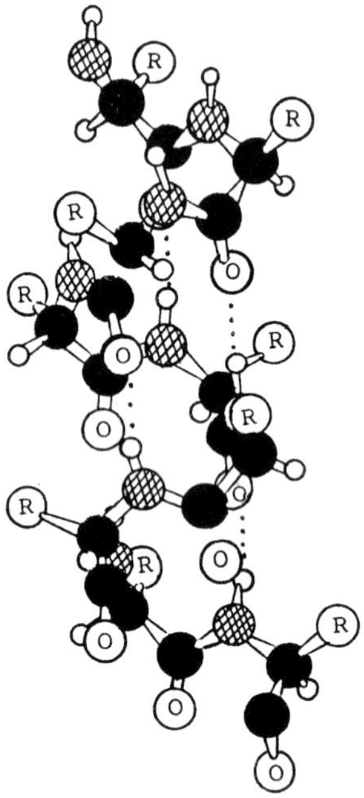

Fig. 1. A fragment of polypeptide chain. It consists of a "skeleton", having a regular structure, and side chains R_1, R_2, R_3, ... In alpha-helical proteins the NH-group of the n-th residue is connected with the CO-group of the (n-4)-th residue (the connection is shown by dotted line).

Fig. 2. The model of the right-handed alpha-helix. Conventional notations: ◍ is the atom of nitrogen (N), ● is the atom of carbon (C), Ⓞ is the atom of oxygen (O), Ⓡ is the amino acid residue, ○ is the atom of hydrogen (H), ... is the hydrogen bond.

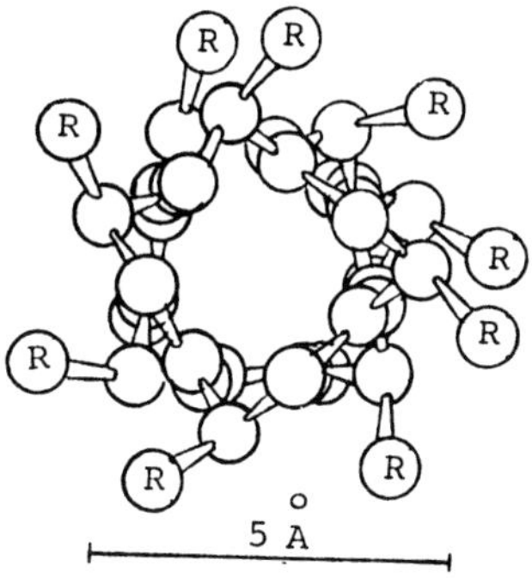

Fig. 3. A scheme of cross-section of the alpha-helix. The side chains (R) are outside. In real systems the Van-der-Waal's radius of the atoms is much more than that shown here. So, there is practically no free space inside the helix.

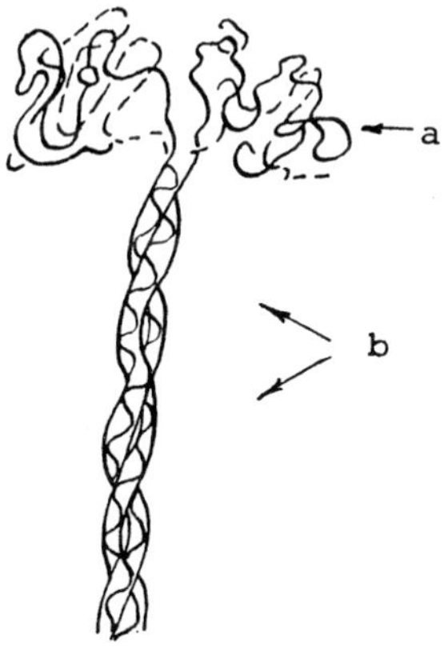

Fig. 4. Myosin molecule has a long "tail" (b), consisting of two superspiralized alpha-helical polypeptide chains. The "head" of the molecule contains four light chains (a).

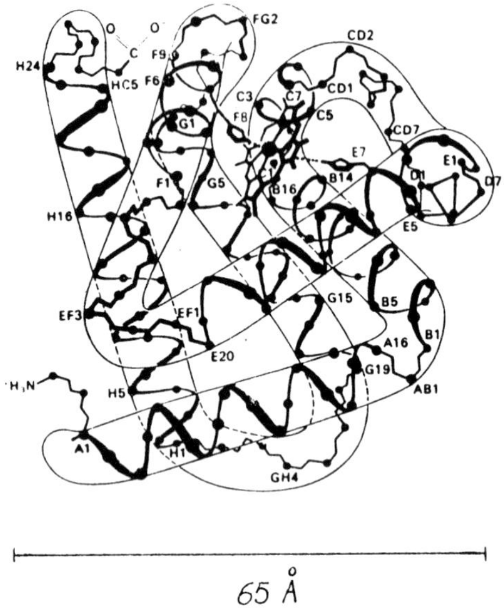

Fig. 5. The model of myoglobin molecule. Only alpha-carbon atoms are shown.

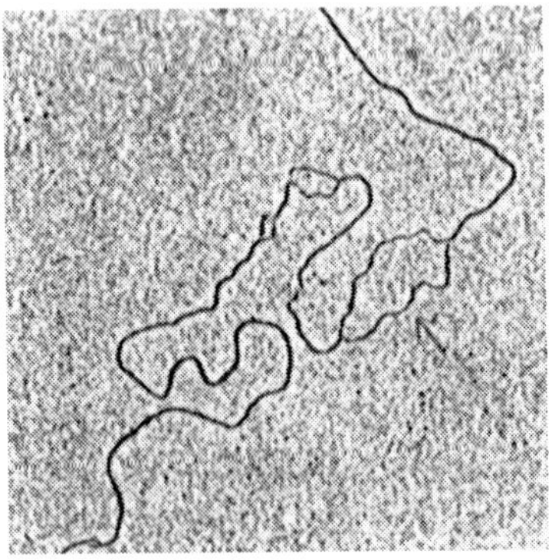

Fig. 6. Electronic microphotography of the DNA molecule.

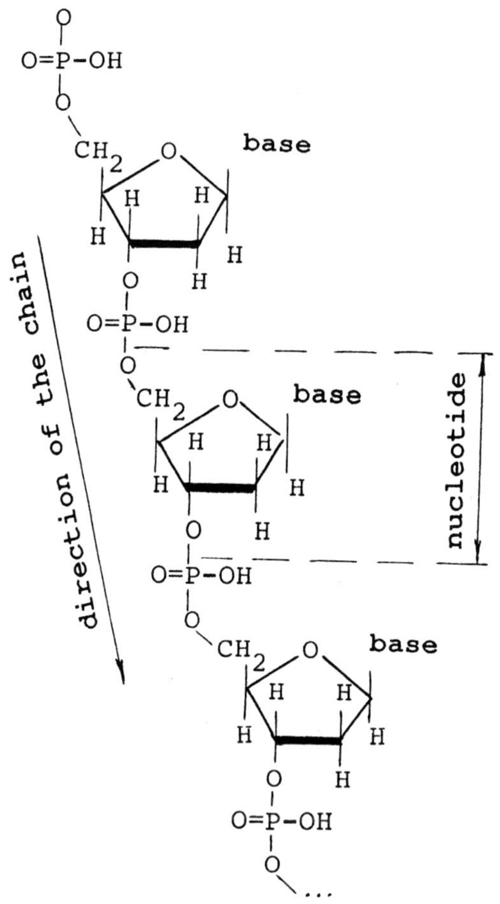

Fig. 7. Fragment of polynucleotide chain.

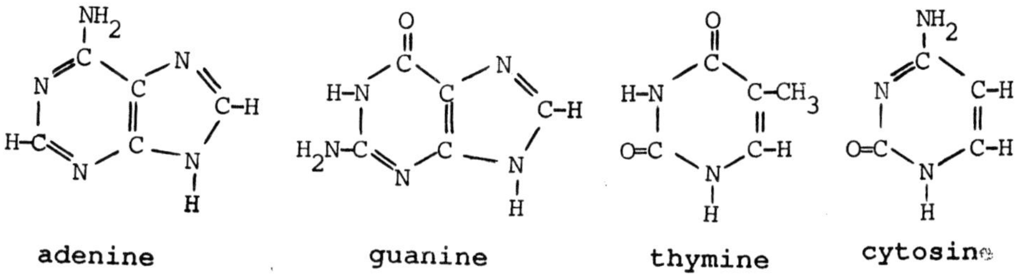

Fig. 8. Nitrous bases which are the structural elements of DNA: adenine, guanine, thymine and cytosine.

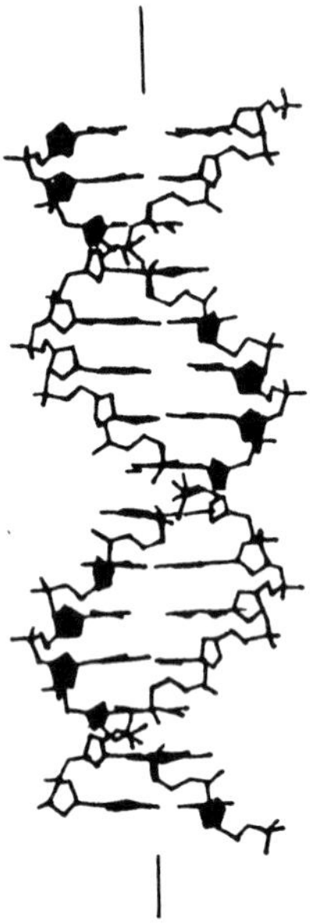

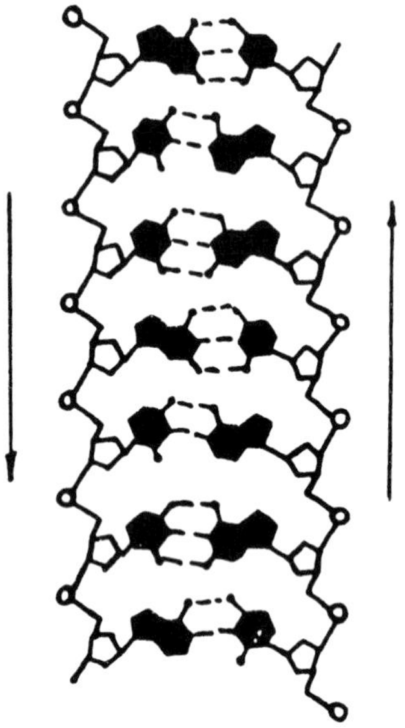

Fig. 9. Schematic picture of a fragment of the DNA double helix. Sugars of one of two polynucleotide chains are painted by black color.

Fig. 10. A scheme of the internal DNA structure. Nitrous bases are painted by black color. Hydrogen bonds are shown by dotted lines.

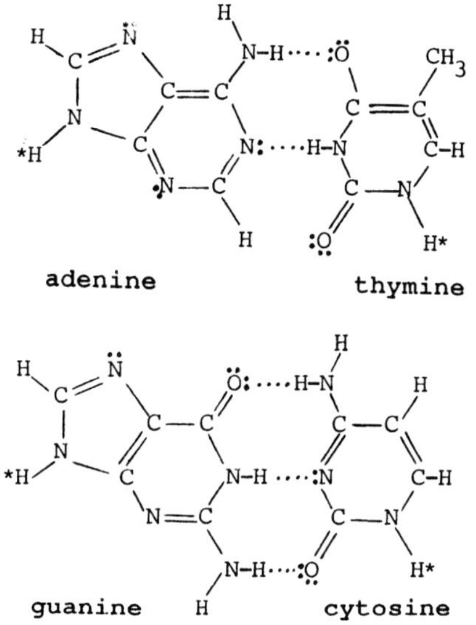

Fig. 11. Pairs of DNA nitrous bases. Hydrogen atoms which are substituted for carbon atoms of sugar rings are indicated by asterisks. Hydrogen bonds are shown by points.

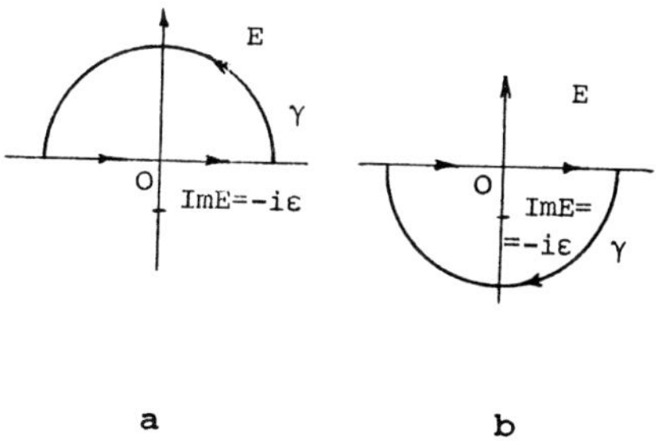

Fig. 12. Energy levels of an isolated molecule.

Fig. 13. Contour of integration (γ) in formula (9-13) for a) t<0 and b) t>0.

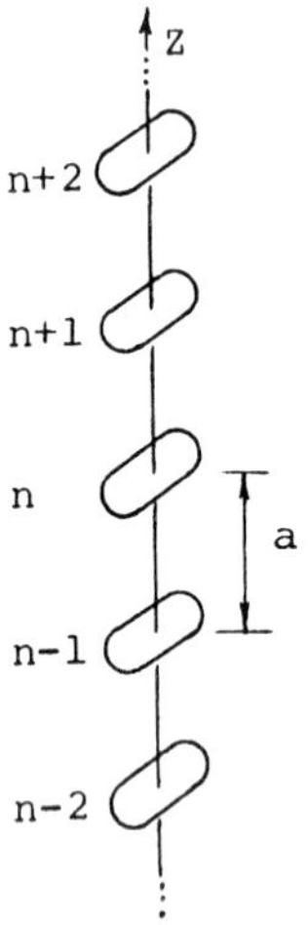

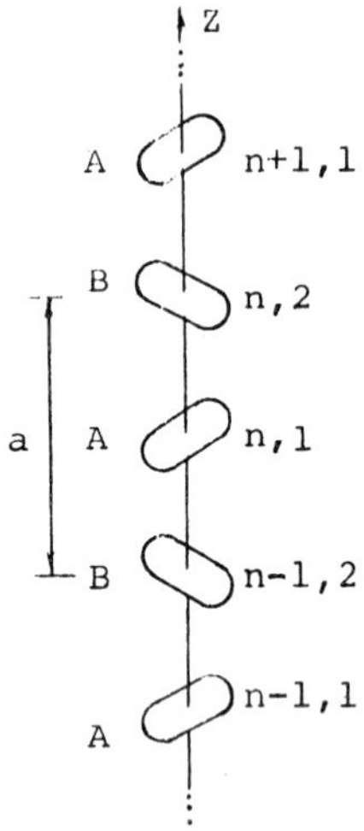

Fig. 14. The chain of identical molecules: n is the number of a molecule (n=1,2,...N); a is the lattice constant.

Fig. 15. The chain of two types the molecules. The pair of numbers (n,α) numerates molecules ($n=1,2,...N$; $\alpha=1,2$); a is the lattice constant.

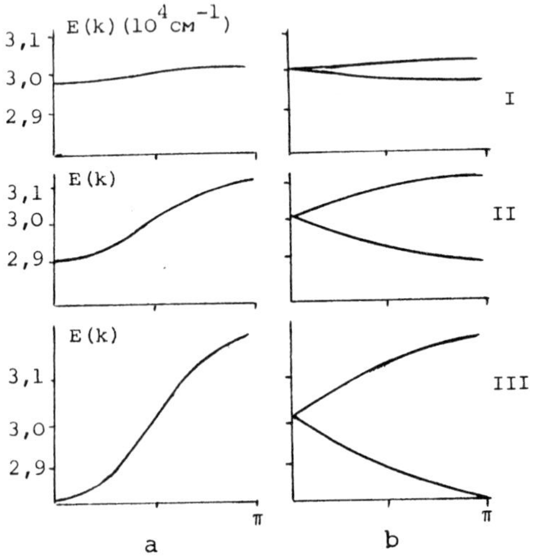

Fig. 16. Dispersion curves for the chain with a) one and b) two molecules in the lattice cell.

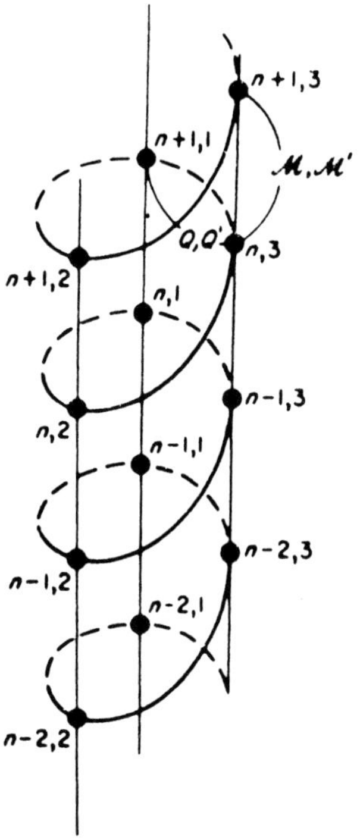

Fig. 17. Simplified model of alpha-helical fragment of protein molecules, proposed by Davydov and Suprun.

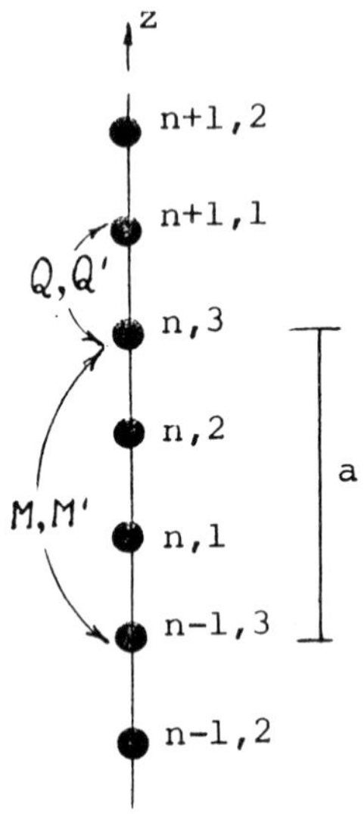

Fig. 18. Molecular chain with three molecules in the lattice cell. The coefficients Q, Q', M and M' characterize interactions with the nearest and the third neighbours; a is the lattice constant.

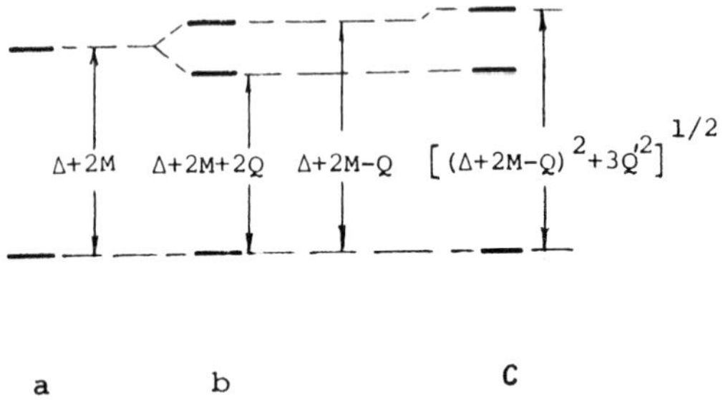

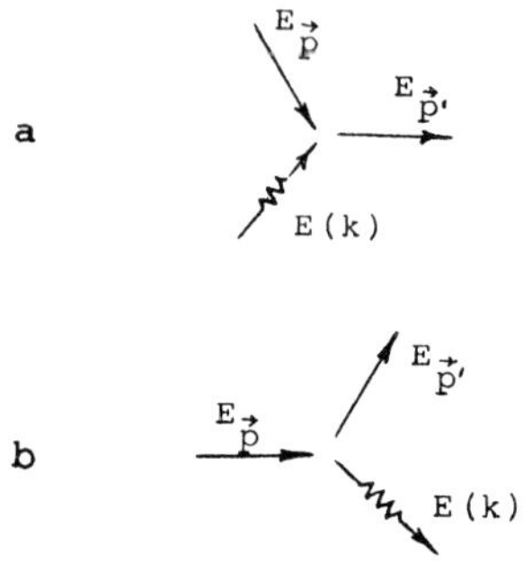

Fig. 19. The arrangement of the energy levels in the model of Davydov and Suprun, calculated a) when interaction between the third neighbours is neglected ($Q=Q'=0$), b) when $Q'=m'=0$ (the Haitler-London approximation) and c) in the general case.

Fig. 20. Annihilation (a) and creation (b) of elementary excitation

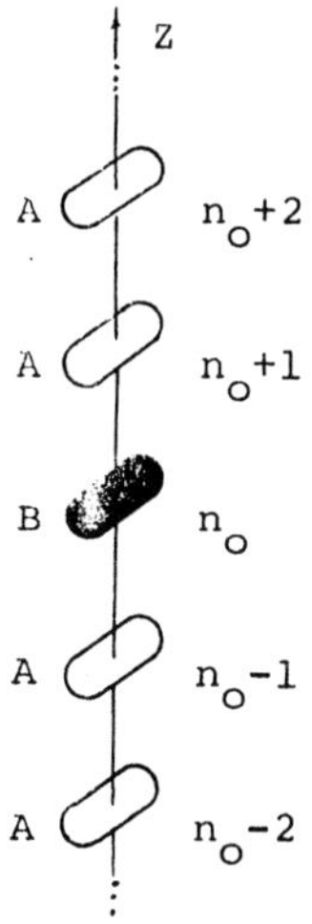

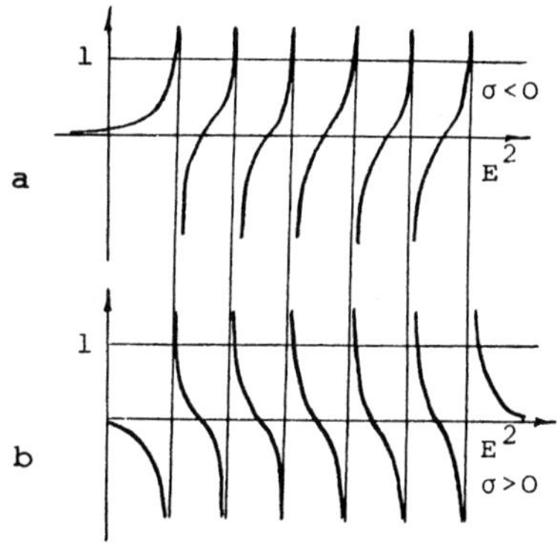

Fig. 21. Molecular chain with one substituted molecule.

Fig. 22. Schematic picture of the graphical solution of eq. (14-15) when a) $\sigma < 0$, b) $\sigma > 0$.

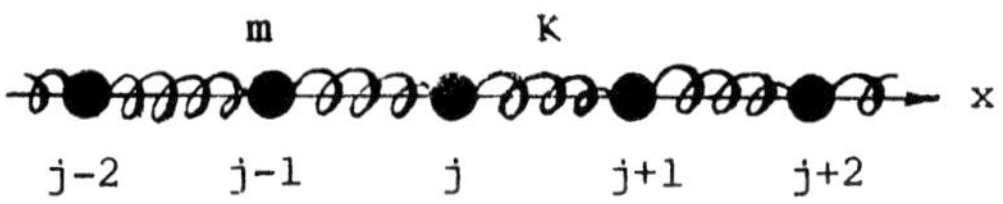

Fig. 23. Discrete analog of the model of elastic rod: the chain consisting of "molecules" connected with one another by springs; m is the mass of the "molecules", k is the rigidity of springs.

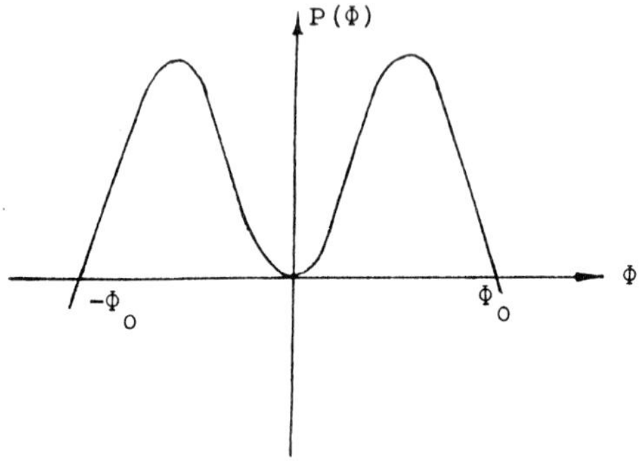

Fig. 24. Schematic picture of the function P(Φ) which are determined by equation (18-23).

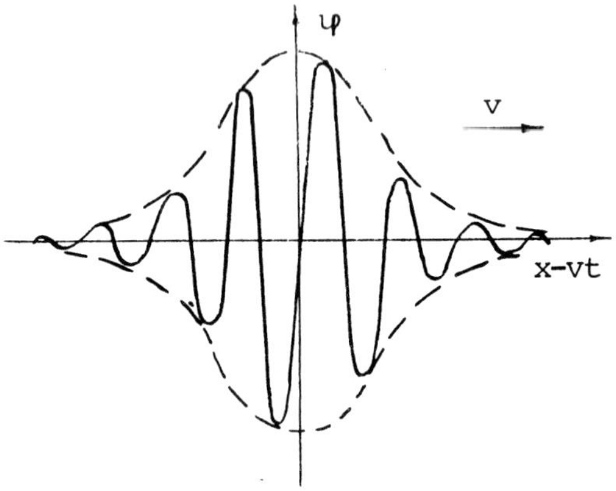

Fig. 25. One-soliton solution of the nonlinear Schrodinger equation.

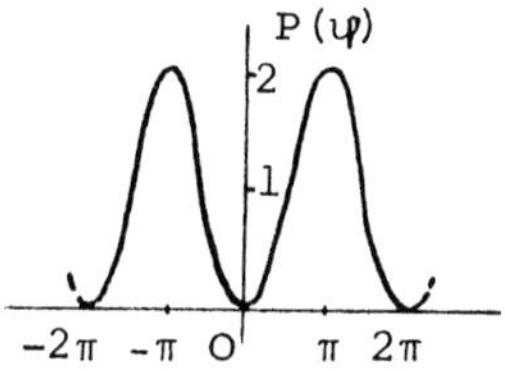

Fig. 26. Schematic picture of the function $P(\varphi)=2(1-\cos\varphi)$.

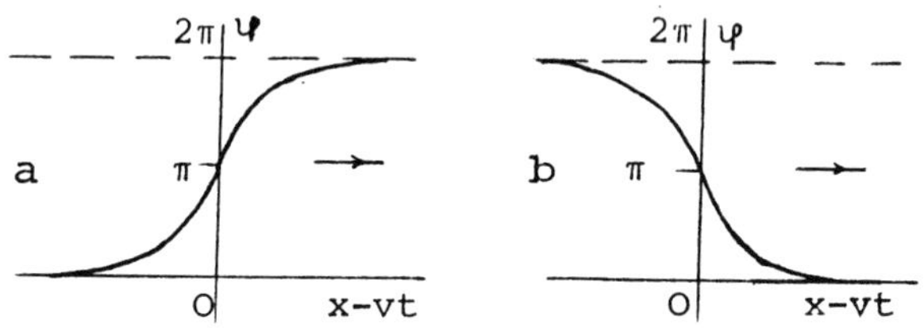

Fig. 27. Kink (a) and antikink (b) of the sine-Gordon equation.

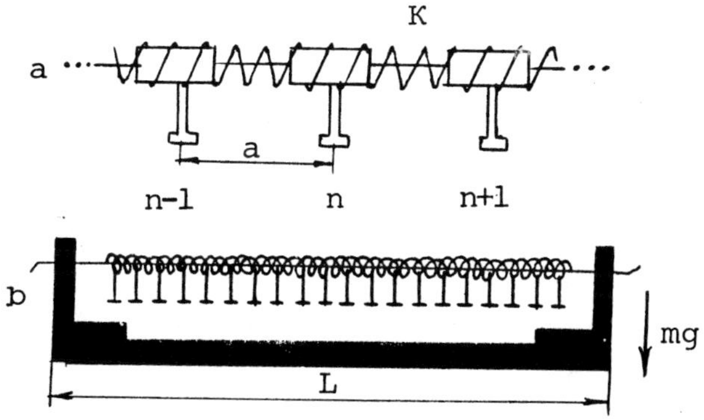

Fig. 28. Mechanical analog of the sine-Gordon equation: (a) a fragment of the chain of pendula, (b) a general view of the model.

Fig. 29. Propagation of a solitary wave in the chain of pendula.

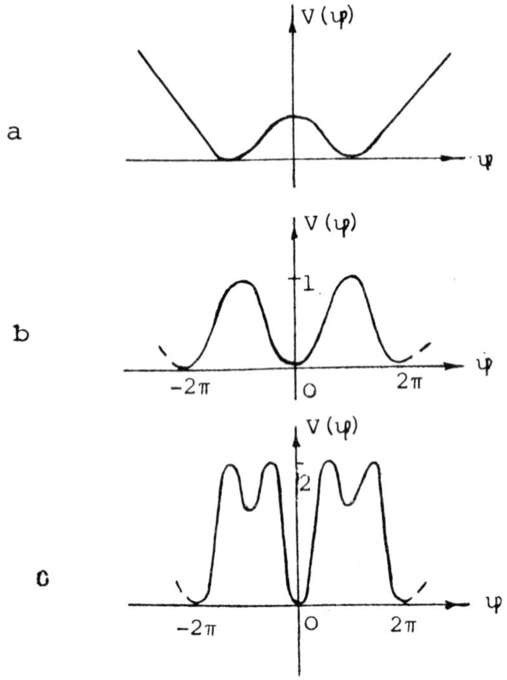

Fig. 30. Potential function V(φ) for a) equation of φ^4, b) sG equation and c) dsG equation.

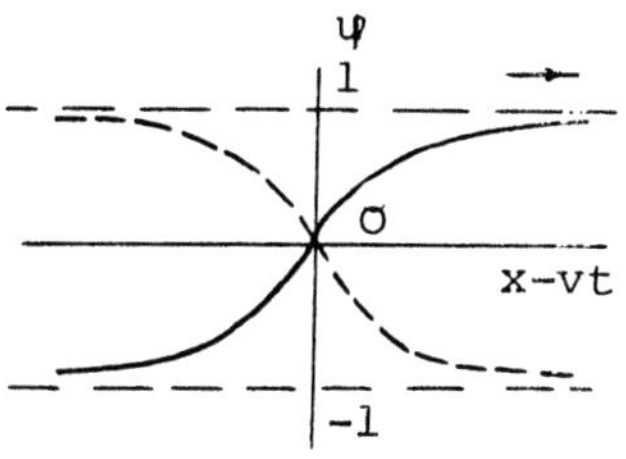

Fig. 31. Kink and antikink of the equation of φ^4.

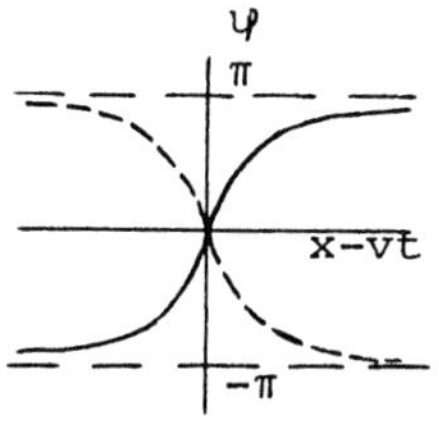

Fig. 32. Kink and antikink of the dsG equation.

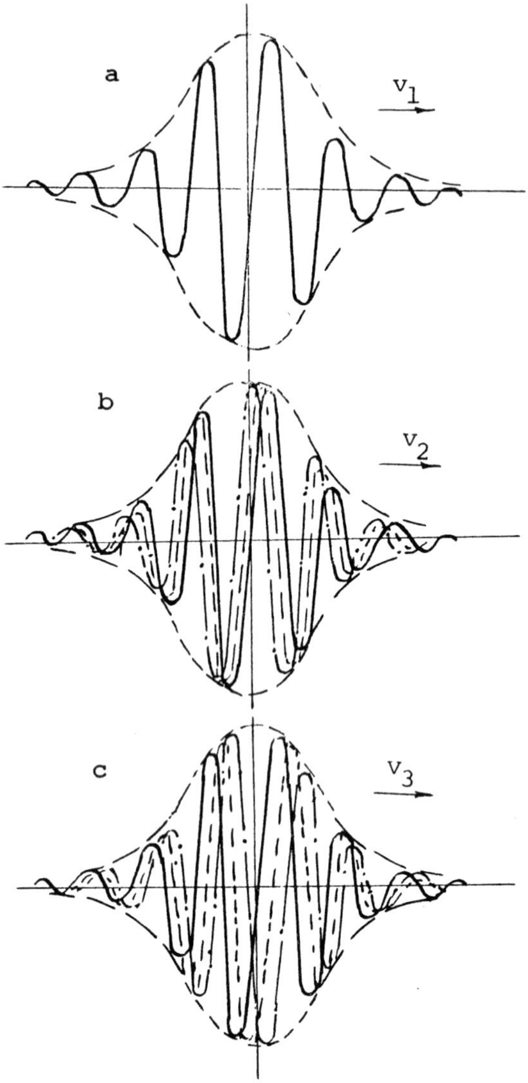

Fig. 33. Three types of soliton-like solutions: modulated waves oscillating (a) in phase, (b) out of phase by $2\pi/3$, (c) out of phase by $4\pi/3$.

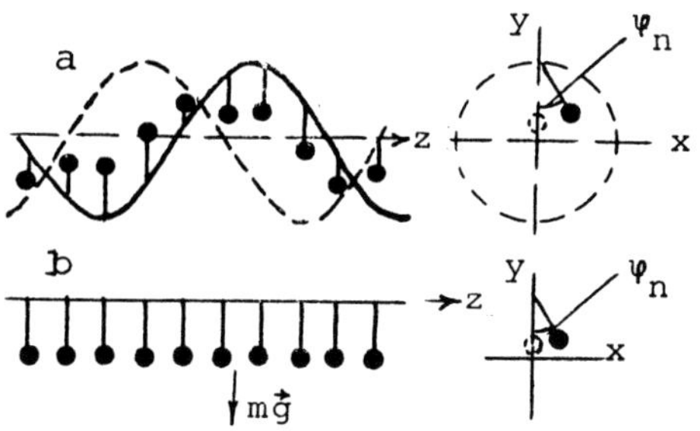

Fig. 34. Schematic picture of (a) the model of DNA and (b) the mechanical analog of the sG equation.

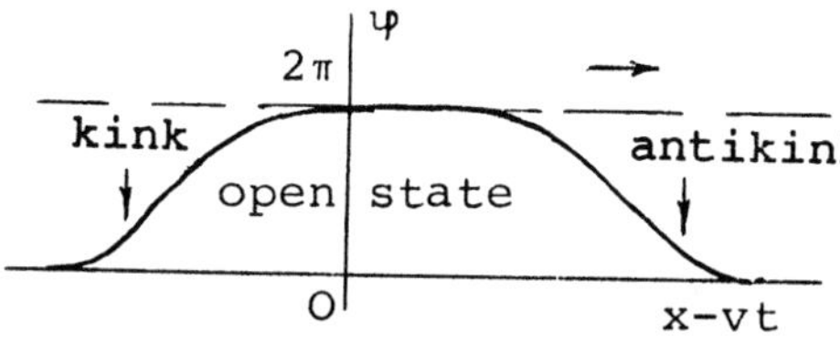

Fig. 35. Pair of kink-antikink describing an open state in DNA.

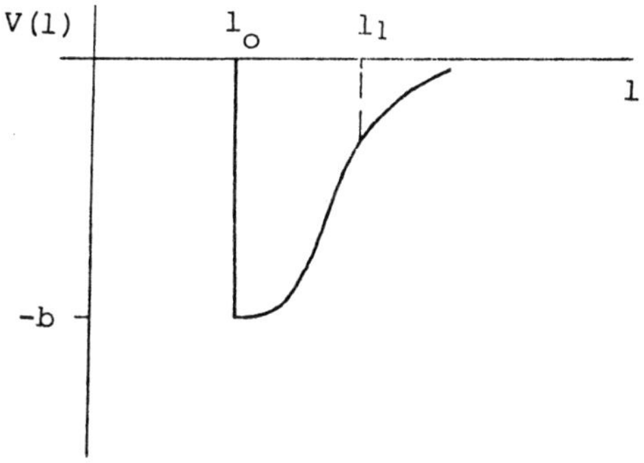

Fig. 36. Schematic picture of the potential function V in the G-L-F model. The dotted line denotes a break of hydrogen bonds between the bases in a pair.

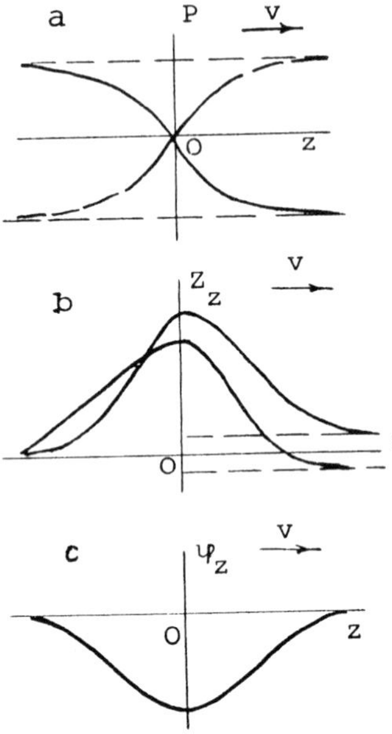

Fig. 37. Schematic picture of the soliton-like solutions of the equations of Krumhansl and Alexander.

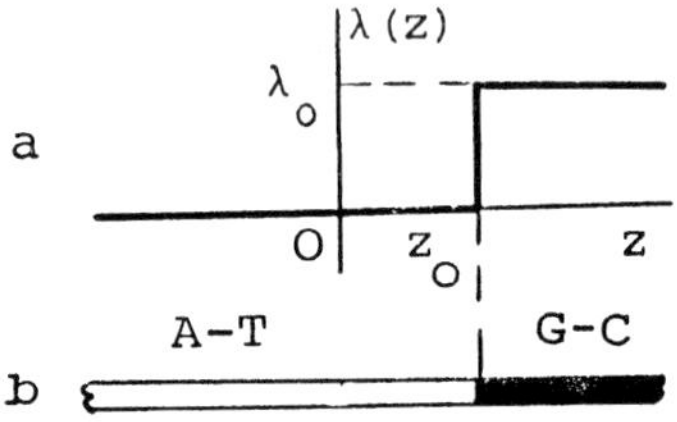

Fig. 38. Schematic picture of (a) the function $\lambda(z)$ simulating the inhomogeneity of the type of boundary between two regions; (b) two regions of DNA: the first one is saturated by A-T pairs, the second one by G-C pairs.

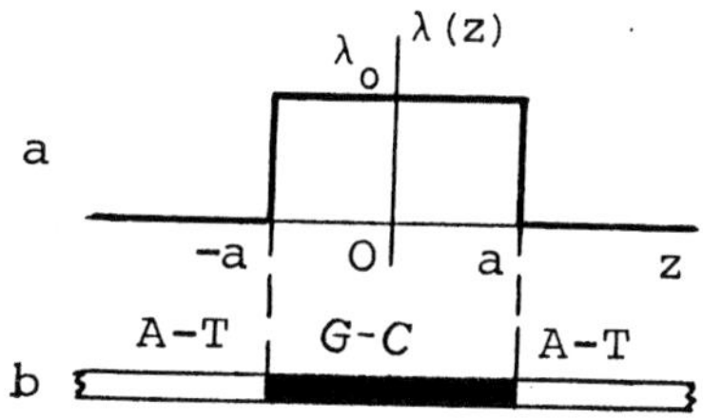

Fig. 39. Schematic picture of (a) the function $\lambda(z)=|\lambda_0|\theta(z-z_0)$ simulating the inhomogeneity of the type of the range of finite length; (b) the G-C block standing out against the rest of the double chain saturated by A-T pairs.

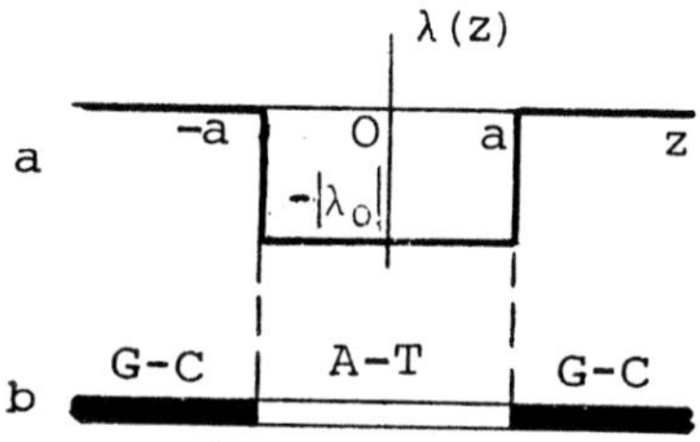

Fig. 40. Schematic picture of (a) the function simulating the inhomogeneity of the type of the range of finite length, (b) the A-T block standing out against the rest of the double chain saturated by A-T pairs.

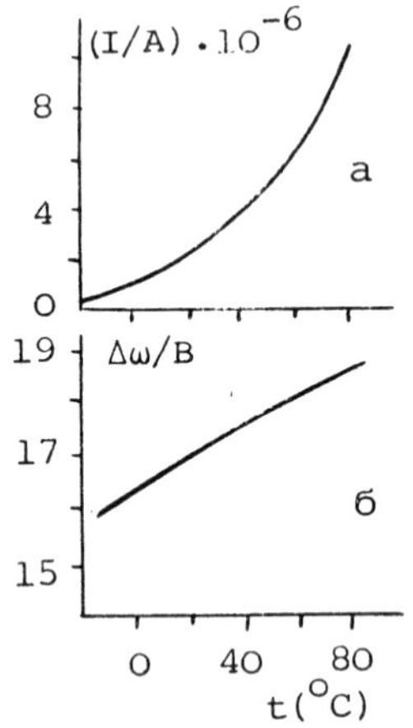

Fig. 41. Graphical picture of the temperature dependence of the integral intensity I and the line width Δω of the central peak in the spectrum of neutron scattering by solitons of DNA.

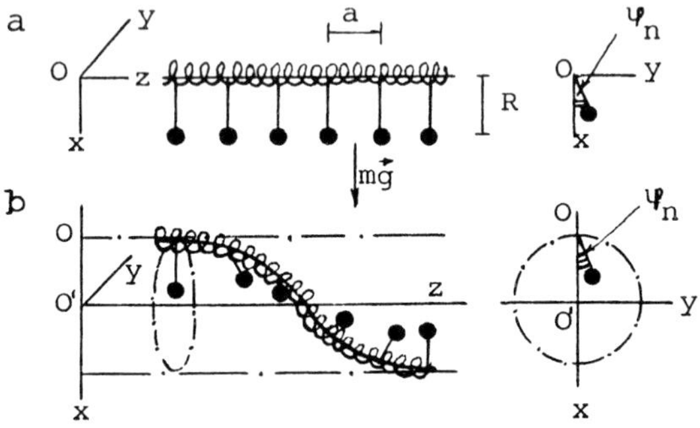

Fig. 42. "Mechanical" models of DNA: a) the linear model, b) the helical model.

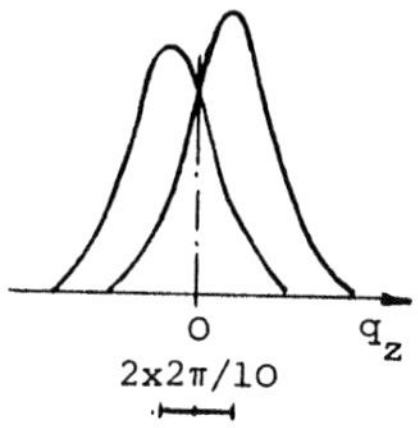

Fig. 43. Schematic picture of the central peak splitting in the spectrum of neutron scattering.

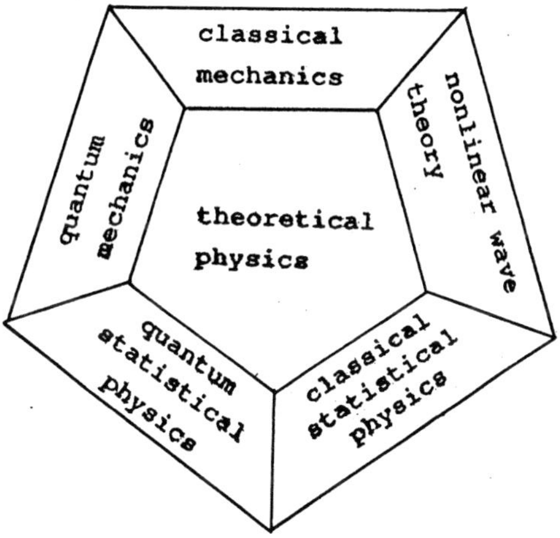

Fig. 44. Diagram of the fields of theoretical physics considered in the book.

REFERENCES

1. Schrödinger E. What is life? The Physical Aspect of the Living Cell. Cambridge University Press, 1955.
2. Szent-Györgyi A. Bioenergetics. Academic Press INC, New York, 1957.
3. Pauling L., Pauling P. Chemistry. W.H. Freeman and Company, San Francisco, 1975.
4. Volkenstein M.V. Molecular Biophysics. Nauka, Moscow, 1975.
5. Stryer L. Biochemistry. W.H. Freeman and Company, San Francisco, 1980.
6. Cantor C.R., Schimmel P.R. Biophysical Chemistry. W.H. Freeman and Company, San Francisco, 1980.
7. Saenger W. Principles of Nucleic Acid Structure. Springer--Verlag, New York - Berlin - Heidelberg - Tokyo, 1984.
8. McCammon J.A., Harvey S.C. Dynamics of Proteins and Nucleic Acids. Cambridge University Press, Cambridge, 1987.
9. Pauling L., Corey R.B. Two hydrogen-bonded spiral configurations of the polypeptide chain. J. Am. Chem. Soc., 72, N11, 5349, 1950.
10. Pauling L., Corey R.B. The structure of proteins: two hydrogen--bonded helical configurations of the polypeptide chain. Proc. Natl. Acad. Sci. USA, 37, N3, 205-211, 1951.
11. Franklin R.E., Gosling R.G. Molecular structure of nucleic acids. Molecular configuration in sodium thymonucleate. Nature, 171, N4356, 740-741, 1953.
12. Watson J.D., Crick F.H.C. Molecular structure of nucleic acids. A structure for deoxyribose nucleic acid. Nature, 171, N4356,

737-738, 1953.

13. Watson J. D., Crick F. H. C. Genetical implications of the structure of deoxyribonucleic acid. Nature, 171, N4361, 964-967, 1953.

14. Wilkins M. H. F., Seeds W. E., Stokes A. R., Wilson H. R. Helical structure of crystalline deoxypentose nucleic acid. Nature, 172, N4382, 759-762, 1953.

15. Bunn Ch. Crystals. Their Role in Nature and in Science. Academic Press, New York-London, 1964.

16. Bogolubov N. N. Lectures on quantum statistics. In: Collection of the Scientific Works. Naukova Dumka, Kiev, 1970, v. 2, 287-350.

17. Tyablikov S. V. Methods of Quantum Theory of Magnetism. Nauka, Moscow, 1975.

18. Zubarev D. N. Green two-time functions in statistical physics. Uspekhy Fizich. Nauk (USSR), 71, N1, 71-116, 1960.

19. Goldstein H. Classical Mechanics. Addison-Wesley Press INC., Cambridge, 1950.

20. Agranovich V. M. Theory of Excitons. Nauka, Moscow, 1968.

21. Davydov A. S. Theory of Excitons. Nauka, Moscow, 1968.

22. Holstein T., Primakoff H. Field dependence of the intrinsic domain magnetization of a ferromagnet. Phys. Rev., 58, N12, 1098-1113, 1940.

23. Chesnut D. B., Suna A. Fermion behavior or one-dimensional excitons. J. Chem. Phys., 39, N1, 146-149, 1963.

24. Schweber S. S. An Introduction to Relativistic Quantum Field Theory. Elmsford, New York, 1961.

25. Bogolubov N. N. Problems of dynamical theory in statistical

physics. In: Collection of the Scientific Works. Naukova Dumka, Kiev, 1970, v. 2, 99-196.

26. Fedyanin V. K., Yakushevich L. V. On the exciton hamiltonian diagonalization. Preprint P17-9627, JINR, Dubna, 1976.

27. Fedyanin V. K. Yakushevich L. V. The spectrum of excitons in a one-dimensional chain with fully orientationally-ordered molecules. Preprint P17-9896, JINR, Dubna, 1976.

28. Fedyanin V. K., Yakushevich L. V. To the problem of exciton--phonon interaction in one-dimensional molecular crystals. I. Linear approximation. Teor. Mat. Fiz. (USSR), 30, N1, 133-137, 1977.

29. Fedyanin V. K., Yakushevich L. V. Elementary excitations in one-dimensional systems with resonance interaction. Teor. Mat. Fiz. (USSR), 37, N3, 371-381, 1978.

30. Fedyanin V. K., Yakushevich L. V. Neutron scattering and exciton spectrum due to impurities in the molecular crystals with the resonance interaction. Solid State Commun., 27, N4, 383-387, 1978.

31. Fedyanin V. K., Yakushevich L. V. The exciton spectrum of the alpha-helical protein molecule model. J. Theor. Biol., 91, N1, 1-11, 1981.

32. Davydov A. C., Suprun A. D. Configurational changes and optical properties of alpha-helical protein molecules. Ukr. Fiz. Z. (USSR), 19, N1, 44-50, 1974.

33. Van Hove L. Correlations in space and time and Born approximation scattering in systems of interacting particles. Phys. Rev., 95, N1, 249-262, 1954.

34. Fermi E. Un generator artificiale di neutroni. Ric. Sci., 7(II),

13–52, 1936.

35. Lifshitz J.M. Optical behaviour of nonideal crystal lattices in the infrared range. I. ZETF (USSR), 12, N3–4, 117–136, 1942.

36. Lifshitz J.M. Optical behaviour of nonideal crystal lattices in the infrared range. II. ZETF (USSR), 12, N3–4, 137–155, 1942.

37. Lifshitz J.M. On degenerate regular perturbations. I. Discrete spectrum. ZETF (USSR), 17, N11, 1017–1025, 1947.

38. Lifshitz J.M. On degenerate regular perturbations. II. Quasi–continuous and continuous spectra. ZETF (USSR), 17, N12, 1076–1089, 1947.

39. Dawber P.G., Elliot R.J. The vibrations of an atom of different mass in a cubic crystal. Proc.Roy.Soc, A–273, N1353, 222–236, 1963.

40. Maradudin A.A. Phonons and lattice imperfections. In: Phonons and Phonon Interactions. Eds. Bak Th.A., Benjamin W.A. INC, New York–Amsterdam, 1964, 424–504.

41. Montroll E.W., Potts R.B. Effects of defects on lattice vibrations. Phys.Rev., 100, N2, 525–543, 1955.

42. Mattis D.C. The Theory of Magnetism. Harper and Row Publishers, New York–Evanston–London, 1965.

43. Scott A.C., Chu F.Y., McLaughlin D.W. The soliton: a new concept in applied science. Proc.IEEE, 61, N10, 1443–1483, 1973.

44. Makhankov V.G., Fedyanin V.K. Nonlinear effects in quasi-one–dimensional models of condensed matter theory. Phys.Reports, 104, N1, 3–86, 1984.

45. Solitons in Action. Eds. Lonngren K., Scott A. Academic Press, New York–San Francisco–London, 1978.

46. Solitons. Eds. Bullough R.K., Candrey P.J. Springer-Verlag, Berlin-Heidelberg-New York, 1980.

47. Calogero F., Degasperis A. Spectral Transform and Solitons. North-Holland Publishing Company, Amsterdam-New York-Oxford, 1982.

48. Whitham G.B. Linear and Nonlinear Waves. John Wiley & Sons, New York-London-Sydney-Toronto, 1974.

49. Fedyanin V.K., Makhankov V.G., Yakushevich L.V. Exciton-phonon interaction in long-range approximation. Phys.Letters, A-61, N4, 256-258, 1977.

50. Fedyanin V.K. Particle-like excitations in the systems described by the Hubbard model without chirality condition. Preprint E-17-12836, JINR, Dubna, 1979.

51. Linder U.,Fedyanin V.K. Solitons in a one-dimensional modified Hubbard model. Phys.stat.sol.(b), 89, N1, 123-129, 1978.

52. Osipov V.A. Fedyanin V.K. The report making at the Third Workshop "Soliton Theory and Applications", Pushchino, 1987.

53. Fedyanin V.K., Yakushevich L.V. Energy migration in one-dimensional structures with resonance interaction. Int.J.Quant.Chem., 16, N2, 215-219, 1979.

54. Fedyanin V.K., Yakushevich L.V. Particle-like excitations in the polypeptide chain model. Int.J.Quant.Chem, 21, N6, 1019-1028, 1982.

55. Zakharov V.E., Shabat A.B. Exact theory of two-dimensional self-focussing and one-dimensional self-modulation of waves in nonlinear media. ZETF (USSR), 61, N10, 1443-1483, 1973.

56. Scott A.C. A nonlinear Klein-Gordon equation. Am.J.Phys., 37, N1, 52-61, 1969.

57. Condat C.A., Guyer R.A., Miller M.D. Double sine-Gordon chain. Phys. Rev., B-27, N1, 474-494, 1983.

58. Landau L.D., Lifshitz E.M. Quantum Mechanics. Fiz.Mat.Iz., Moscow, 1963.

59. Rubinstein J. Sine-Gordon equation. J.Math.Phys., 11, N1, 258-266, 1970.

60. Fedyanin V.K., Yakushevich L.V. Scattering of neutrons and light by DNA solitons. Studia biophys., 103, N3, 171-178, 1984.

61. Yakushevich L.V. The effects of damping, external fields and inhomogeneity on the nonlinear dynamics of biopolymers. Studia biophys., 121, N3, 201-207, 1987.

62. Yakushevich L.V. Nonlinear DNA dynamics: a new model. Phys. Lett., A-136, N7-8, 413-417, 1989.

63. Davydov A.S. Solitons in quasi-one-dimensional molecular structures. Usp.Fiz.Nauk (USSR), 138, N4, 603-642, 1982.

64. Yakushevich L.V. Elementary excitations in one-dimensional systems with resonance interaction. Doct. thesis, JINR, Dubna, 1978.

65. Makhankov V.G., Fedyanin V.K. Soliton-like solutions of the equations describing excitations in one-dimensional molecular crystals. Dokl.Acad.Nauk USSR, 236, N4, 838-841, 1977.

66. Englander S.W., Kallenbach N.R., Heeger A.J., Krumhansl J.A., Litwin S. Nature of the open state in long polynucleotide double helices: possibility of soliton excitations. Proc.Natl. Acad.Sci.USA, 77, N12, 7222-7226, 1980.

67. Sobell H.M. Kink-antikink bound states in DNA structures. In: Biological Macromolecules and Assemblies. V.2. Nucleic Acids and Interactive Proteins. Eds. Jurnak F.A., McPherson A. John

Wiley and Sons, New York, 1984, 171–232.

68. Fedyanin V.K., Gochev I., Lisy V. Nonlinear dynamics of bases in a continual model of DNA double helices. Studia biophys., 116, N1, 59–64, 1986.

69. Fedyanin V.K., Lisy V. Soliton conformational excitations in DNA. Studia biophys., 116, N1, 65–71, 1986.

70. Yomosa S. Solitary excitations in deoxyribonucleic acid (DNA) double helices. Phys. Rev., A-30, N1, 474–480, 1984.

71. Takeno S., Homma S. Topological solitons and modulated structure of bases in DNA double helices. Prog. Theor. Phys., 70, N1, 308–311, 1983.

72. Homma S., Takeno S. A coupled base-rotator model for structure and dynamics of DNA. Prog. Theor. Phys., 72, N4, 679–693, 1984.

73. Zhang Ch.-T. Soliton excitations in deoxyribonucleic acid (DNA) double helices. Phys. Rev., A-35, N2, 886–891, 1987.

74. Volkov S.N. Nonlinear waves and conformational mobility of DNA. Preprint ITP-84-52P, Kiev, 1984.

75. Volkov S.N. Conformational transition dynamics and mechanism of long range effects in DNA. J. Theor. Biol., 143, N4, 485–496, 1990.

76. Krumhansl J.A., Alexander D.M. Nonlinear dynamics and conformational excitations in biomolecular materials. In: Structure and Dynamics: Nucleic Acids and Proteins. Eds. Clementi E., Sarma R.H. Adenine Press, New York, 1983, 61–80.

77. Mikeska H.J. Solitons in a one-dimensional magnet with an easy plane. J. Phys. C.: Solid State Phys., 11, N1, L-29–L-32, 1978.

78. Boucher J.P., Regnault L.P., Rossat-Mignod J., Renard J.P., Bouillot J., Stirling W.G. Experimental evidence for magnetic

solitons in the antiferromagnetic chains of $(CD_3)_4NMnCl_3$. J. Appl. Phys., 52, N3, part 2, 1956–1960, 1981.

79. Kjems J. K., Steiner M. Evidence for soliton modes in the one-dimensional ferromagnet $CsNiF_3$. Phys. Rev. Lett. 41, N16, 1137–1140, 1978.

80. Marshall W., Lovesey S. W. Theory of Thermal Neutron Scattering. University Press, Oxford, 1971.

81. Hakim M. B., Lindsay S. M., Powell J. The speed of sound in DNA. Biopolymers, 23, N7, 1185–1192, 1984.

82. Komarov L. I., Fisher I. Z. On the theory of Rayleigh scattering of light by liquids. ZETF (USSR), 43, N5(11), 1927–1933, 1962.

83. Pecora R. Doppler shifts in light scattering. II. Flexible polymer molecules. J. Chem. Phys., 43, N5, 1562–1564, 1965.

84. Gaeta G. On a model of DNA torsion dynamics. Phys. Lett., A-143, N4-5, 227-232, 1990.

SUBJECT INDEX